HOLT SCIENCE & TECHNOLOGY

Life Science

Study Guide

HOLT, RINEHART AND WINSTON

A Harcourt Education Company

Orlando • Austin • New York • San Diego • Toronto • London

TO THE STUDENT

Do you need to practice for an upcoming section quiz or chapter test? If so, this booklet will help you. The *Study Guide* is a tool that allows you to confirm what you know and to identify topics you do not understand so that you can succeed in your studies. These worksheets are reproductions of each Section Review and Chapter Review in your textbook, but there is one difference—the *Study Guide* worksheets provide plenty of space for you to record your answers and write your thoughts and ideas.

Use these worksheets in the following ways:

• as a learning tool to work interactively with the textbook by answering the questions as you read the text

• as a review to test your understanding of the main concepts and terminology

• as a practice quiz or test to prepare for a section quiz or chapter test

ISBN 978-0-03-045562-9   ISBN 0-03-045562-6
14 1409 13
4500444417

Contents

The World of Life Science

Section Review Worksheets

It's Alive!! Or Is It?

Section Review Worksheets

Cells: The Basic Units of Life

Section Review Worksheets

The Cell in Action

Section Review Worksheets

Heredity

Section Review Worksheets

Genes and DNA

Section Review Worksheets

Chapter Review Worksheets

The Evolution of Living Things

Section Review Worksheets

Chapter Review Worksheets

The History of Life on Earth

Section Review Worksheets

Chapter Review Worksheets

Classification

Section Review Worksheets

Chapter Review Worksheets

Bacteria and Viruses

Section Review Worksheets

Chapter Review Worksheets

Protists and Fungi

Section Review Worksheets

Chapter Review Worksheets

Introduction to Plants

Section Review Worksheets

Plant Processes

Section Review Worksheets

Animals and Behavior

Section Review Worksheets

Invertebrates

Section Review Worksheets

Fishes, Amphibians, and Reptiles

Section Review Worksheets

Birds and Mammals

Section Review Worksheets

Interactions of Living Things

Section Review Worksheets

Cycles in Nature

Section Review Worksheets

The Earth's Ecosystems

Section Review Worksheets

Environmental Problems and Solutions

Section Review Worksheets

Body Organization and Structure

Section Review Worksheets

Chapter Review Worksheets

Circulation and Respiration

Section Review Worksheets

Chapter Review Worksheets

The Digestive and Urinary Systems

Section Review Worksheets

Chapter Review Worksheets

Communication and Control

Section Review Worksheets

Chapter Review Worksheets

Reproduction and Development

Section Review Worksheets

Chapter Review Worksheets

Body Defenses and Disease

Section Review Worksheets

Chapter Review Worksheets

Staying Healthy

Section Review Worksheets

Chapter Review Worksheets

Name _____ Class _____ Date _____

Skills Worksheet

Section Review

Asking About Life
USING KEY TERMS

1. In your own words, write a definition for the term *life science*.

_____

UNDERSTANDING KEY IDEAS

_____ **2.** Life scientists may study any of the following EXCEPT
   a. things that were once living.
   b. environmental problems.
   c. stars in outer space.
   d. diseases that are not inherited by humans.

3. What is the importance of asking questions in life science?

_____

_____

_____

4. Where do life scientists work? What do life scientists study?

_____

_____

_____

MATH SKILLS

5. Students in a science class collected 50 frogs from a pond and found that 15 of these frogs had deformities. What percentage of the frogs had deformities? Show your work below.

| Section Review *continued*

CRITICAL THINKING

6. Identifying Relationships Make a list of five things you do or deal with daily. Give an example of how life science might relate to each of these things.

_____

_____

_____

_____

_____

_____

_____

_____

_____

_____

7. Applying Concepts Look at Figure 5. Propose five questions about what you see. Share one of your questions with your classmates.

_____

_____

_____

_____

_____

_____

_____

_____

_____

_____

Skills Worksheet

Section Review

Scientific Methods

USING KEY TERMS

1. Use the following terms in the same sentence: *hypothesis, controlled experiment,* and *variable.*

_____

_____

_____

UNDERSTANDING KEY IDEAS

_____ **2.** The steps of scientific methods
   a. are exactly the same in every investigation.
   b. must always be used in the same order.
   c. are not always used in the same order.
   d. always end with a conclusion.

3. What are the essential parts of a controlled experiment?

_____

_____

4. What causes scientific knowledge to change?

_____

_____

MATH SKILLS

5. Calculate the average of the following values: 4, 5, 6, 6, 9. Show your work below.

Section Review *continued*

CRITICAL THINKING

6. Analyzing Methods Why was UV light chosen to be the variable in the frog experiment?

_____

_____

_____

_____

7. Analyzing Processes Why are there many ways to follow the steps of scientific methods?

_____

_____

_____

8. Making Inferences Why might two scientists working on the same problem draw different conclusions?

_____

_____

_____

_____

9. Identifying Bias Investigations often begin with observation. How does observation limit what scientists can study?

_____

_____

_____

_____

Section Review *continued*

INTERPRETING GRAPHICS

10. The table below shows how long it takes for one bacterium to divide and become two bacteria. Plot this information on a graph, with temperature on the x-axis and the time to double on the y-axis. Do not graph values for which there is no growth. What temperature allows the bacteria to multiply most quickly?

Temperature (°C)	Time to double (minutes)
10	130
20	60
25	40
30	29
37	17
40	19
45	32
50	no growth

Skills Worksheet

Section Review

Scientific Models

USING KEY TERMS

In each of the following sentences, replace the incorrect term with the correct term from the word bank.

theory   law

1. A law is an explanation that matches many hypotheses but may still change.

_____

2. A model tells you exactly what to expect in certain situations.

_____

UNDERSTANDING KEY IDEAS

_____ **3.** A limitation of models is that
   a. they are large enough to see.
   b. they do not act exactly like the things that they model.
   c. they are smaller than the things that they model.
   d. they model unfamiliar things.

4. What are three types of models? Give an example of each type.

_____

_____

_____

5. Compare how scientists use theories with how they use laws.

_____

_____

_____

_____

_____

Section Review *continued*

MATH SKILLS

6. If Jerry is 2.1 m tall, how tall is a scale model of Jerry that is 10% of his size? Show your work below.

CRITICAL THINKING

7. Applying Concepts You and a friend are making a three-dimensional model of an extinct plant. Describe some of the potential uses for your model. What are some limitations of your model?

_____

_____

_____

_____

_____

_____

_____

_____

_____

Name _____ Class _____ Date _____

Section Review

Tools, Measurement, and Safety

USING KEY TERMS

Complete each of the following sentences by choosing the correct term from the word bank.

mass	area
volume	temperature

1. The measure of the surface of an object is called _____.

2. Life scientists use kilograms when measuring an object's

_____.

3. The _____ of a liquid is usually described in liters.

UNDERSTANDING KEY IDEAS

_____ **4.** SI units are
   a. always based on standardized measurements of body parts.
   b. almost always based on the number 10.
   c. used only to measure length.
   d. used only in France.

5. How is temperature related to energy?

_____

_____

6. If you were going to measure the mass of a fly, which SI unit would be most appropriate?

_____

_____

Section Review *continued*

MATH SKILLS

7. Convert 3.0 L into cubic centimeters. Show your work below.

8. Calculate the volume of a textbook that is 28.5 cm long, 22 cm wide, and 3.5 cm thick. Show your work below.

CRITICAL THINKING

9. Making Inferences The mite shown in your textbook is about 500 μm long in real life. What tool was probably used to produce this image? How can you tell?

_____

_____

_____

_____

_____

10. Applying Concepts Give an example of what could happen if you do not follow safety rules.

_____

_____

_____

_____

_____

Skills Worksheet

Chapter Review

USING KEY TERMS

1. Use the following terms in the same sentence: *life science* and *scientific methods.*

_____

_____

_____

2. Use the following terms in the same sentence: *controlled experiment* and *variable.*

_____

_____

_____

For each pair of terms, explain how the meanings of the terms differ.

3. *theory* and *hypothesis*

_____

_____

_____

_____

4. *compound light microscope* and *electron microscope*

_____

_____

_____

_____

5. *area* and *volume*

_____

_____

_____

_____

UNDERSTANDING KEY IDEAS

Multiple Choice

_____ **6.** The steps of scientific methods
  a. must all be used in every scientific investigation.
  b. must always be used in the same order.
  c. often start with a question.
  d. always result in the development of a theory.

Chapter Review *continued*

_____ **7.** In a controlled experiment,
  a. a control group is compared with one or more experimental groups.
  b. there are at least two variables.
  c. all factors should be different.
  d. a variable is not needed.

_____ **8.** Which of the following tools is best for measuring 100 mL of water?
  a. 10 mL graduated cylinder
  b. 150 mL graduated cylinder
  c. 250 mL beaker
  d. 500 mL beaker

_____ **9.** Which of the following is NOT an SI unit?
  a. meter   **c.** liter
  b. foot   **d.** kilogram

_____ **10.** A pencil is 14 cm long. How many millimeters long is it?
  a. 1.4 mm   **c.** 1,400 mm
  b. 140 mm   **d.** 1,400,000 mm

_____ **11.** The directions for a lab include the safety icons shown below. These icons mean that

  a. you should be careful.
  b. you are going into the laboratory.
  c. you should wash your hands first.
  d. you should wear safety goggles, a lab apron, and gloves during the lab.

Short Answer

12. List three ways that science is beneficial to living things.

_____

_____

_____

13. Why do hypotheses need to be testable?

_____

_____

_____

Chapter Review *continued*

14. Give an example of how a life scientist might use computers and technology.

_____

_____

_____

15. List three types of models, and give an example of each.

_____

_____

_____

16. What are some advantages and limitations of models?

_____

_____

_____

17. Which SI units can be used to describe the volume of an object? Which SI units can be used to describe the mass of an object?

_____

_____

_____

18. In a controlled experiment, why should there be several individuals in the control group and in each of the experimental groups?

_____

_____

_____

CRITICAL THINKING

19. Concept Mapping Use the following terms to create a concept map: *observations, predictions, questions, controlled experiments, variable,* and *hypothesis.*

20. Making Inferences Investigations often begin with observation. What limits are there to the observations that scientists can make?

_____

_____

21. Forming Hypotheses A scientist who studies mice observes that on the day the mice are fed vitamins with their meals, they perform better in mazes. What hypothesis would you form to explain this phenomenon? Write a testable prediction based on your hypothesis.

_____

_____

Chapter Review *continued*

INTERPRETING GRAPHICS

The pictures below show how an egg can be measured by using a beaker and water. Use the pictures to answer the questions that follow.

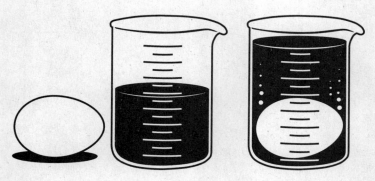

Before: 125 mL   After: 200 mL

_____ **22.** What kind of measurement is being taken?
    a. area   **c.** mass
    b. length   **d.** volume

_____ **23.** Which of the following is an accurate measurement of the egg in the picture?
    a. 75 cm^3 **c.** 125 mL
    b. 125 cm^3 **d.** 200 mL

24. Make a double line graph from the data in the following table.

Number of Frogs		
Date	**Normal**	**Deformed**
1995	25	0
1996	21	0
1997	19	1
1998	20	2
1999	17	3
2000	20	5

Skills Worksheet

Section Review

Characteristics of Living Things

USING KEY TERMS

Complete each of the following sentences by choosing the correct term from the word bank.

cells   stimulus

homeostasis   metabolism

1. Sunlight can be a _____.

2. Living things are made of _____.

UNDERSTANDING KEY IDEAS

_____ **3.** Homeostasis means maintaining
   a. stable internal conditions.
   b. varied internal conditions.
   c. similar offspring.
   d. varied offspring.

4. Explain the difference between asexual and sexual reproduction.

_____

_____

5. Describe the six characteristics of living things.

_____

_____

_____

_____

_____

_____

| Section Review *continued*

MATH SKILLS

6. Bacteria double every generation. One bacterium is in the first generation. How many are in the sixth generation? Show your work below.

CRITICAL THINKING

7. Applying Concepts How do you respond to some stimuli in your environment?

_____

_____

_____

_____

8. Identifying Relationships What does the fur coat of a bear have to do with homeostasis?

_____

_____

_____

_____

Skills Worksheet

Section Review

The Necessities of Life

USING KEY TERMS

For each pair of terms, explain how the meanings of the terms differ.

1. *producer* and *consumer*

_____

_____

2. *lipid* and *phospholipid*

_____

_____

UNDERSTANDING KEY IDEAS

_____ **3.** Plants store extra sugar as
   a. proteins.
   b. starch.
   c. nucleic acids.
   d. phospholipds.

4. Explain why organisms need food, water, air, and living space.

_____

_____

5. Describe the chemical building blocks of cells.

_____

_____

6. Why are decomposers categorized as consumers? How do they differ from producers?

_____

_____

7. What are the subunits of proteins?

_____

_____

| Section Review *continued*

MATH SKILLS

8. Protein A is a chain of 660 amino acids. Protein B is a chain of 11 amino acids. How many times more amino acids does protein A have than protein B? Show your work below.

CRITICAL THINKING

9. Making Inferences Could life as we know it exist on Earth if air contained only oxygen? Explain.

_____

_____

_____

_____

10. Identifying Relationships How might a cave, an ant, and a lake each meet the needs of an organism?

_____

_____

_____

11. Predicting Consequences What would happen to the supply of ATP in your cells if you did not eat enough carbohydrates? How would this affect your cells?

_____

_____

_____

12. Applying Concepts Which resource do you think is most important to your survival: water, air, a place to live, or food? Explain your answer.

_____

_____

_____

_____

Skills Worksheet

Chapter Review

USING KEY TERMS

Complete each of the following sentences by choosing the correct term from the word bank.

lipid	carbohydrate
consumer	heredity
homeostasis	producer

1. The process of maintaining a stable internal environment is known

as _____.

2. Offspring resemble their parents because of _____.

3. A _____ obtains food by eating other organisms.

4. Starch is a _____ and is made up of sugars.

5. Fat is a _____ that stores energy for an organism.

UNDERSTANDING KEY IDEAS

Multiple Choice

_____ **6.** Which of the following statements about cells is true?
  a. Cells are the structures that contain all of the materials necessary for life.
  b. Cells are found in all organisms.
  c. Cells are sometimes specialized for particular functions.
  d. All of the above

_____ **7.** Which of the following statements about all living things is true?
  a. All living things reproduce sexually.
  b. All living things have one or more cells.
  c. All living things must make their own food.
  d. All living things reproduce asexually.

_____ **8.** Organisms must have food because
  a. food is a source of energy.
  b. food supplies cells with oxygen.
  c. organisms never make their own food.
  d. All of the above

_____ **9.** A change in an organism's environment that affects the organism's activities is a
  a. response.   **c.** metabolism.
  b. stimulus.   **d.** producer.

| Chapter Review *continued*

_____10. Organisms store energy in
     a. nucleic acids.   **c.** lipids.
     b. phospholipids.   **d.** water.

_____11. The molecule that contains the information about how to make proteins is
     a. ATP.
     b. a carbohydrate.
     c. DNA.
     d. a phospholipid.

_____12. The subunits of nucleic acids are
     a. nucleotides.
     b. oils.
     c. sugars.
     d. amino acids.

Short Answer

13. What is the difference between asexual reproduction and sexual reproduction?

_____

_____

_____

14. In one or two sentences, explain why living things must have air.

_____

_____

_____

15. What is ATP, and why is it important to a cell?

_____

_____

_____

CRITICAL THINKING

16. Concept Mapping Use the following terms to create a concept map: *cell,
carbohydrates, protein, enzymes, DNA, sugars, lipids, nucleotides, amino
acids,* and *nucleic acid.*

| Chapter Review *continued*

17. Analyzing Ideas A flame can move, grow larger, and give off heat. Is a flame alive? Explain.

_____

_____

18. Applying Concepts Based on what you know about carbohydrates, lipids, and proteins, why is it important for you to eat a balanced diet?

_____

_____

19. Evaluating Hypotheses Your friend tells you that the stimulus of music makes his goldfish swim faster. How would you design a controlled experiment to test your friend's claim?

_____

_____

INTERPRETING GRAPHICS

The pictures below show the same plant over a period of 3 days. Use the pictures below to answer the questions that follow.

Day 1   **Day 2**   **Day 3**

20. What is the plant doing?

_____

_____

21. What characteristic(s) of living things is the plant exhibiting?

_____

_____

Section Review

The Diversity of Cells

USING KEY TERMS

1. In your own words, write a definition for the term *organelle*.

   _____

   _____

2. Use the following terms in the same sentence: *prokaryotic*, *nucleus*, and *eukaryotic*.

   _____

   _____

   _____

UNDERSTANDING KEY IDEAS

_____ 3. Cell size is limited by the
   a. thickness of the cell wall.
   b. size of the cell's nucleus.
   c. cell's surface area–to-volume ratio.
   d. amount of cytoplasm in the cell.

4. What are the three parts of the cell theory?

   _____

   _____

   _____

5. Name three structures that every cell has.

   _____

   _____

   _____

6. Give two ways in which archaea are different from bacteria.

   _____

   _____

Name _____ Class _____ Date _____

CRITICAL THINKING

7. Applying Concepts You have discovered a new single-celled organism. It has a cell wall, ribosomes, and long, circular DNA. Is it a eukaryote or a prokaryote cell? Explain.

_____

_____

8. Identifying Relationships One of your students brings you a cell about the size of the period at the end of this sentence. It is a single cell, but it also forms chains. What characteristics would this cell have if the organism is a eukaryote? If it is a prokaryote? What would you look for first?

_____

_____

INTERPRETING GRAPHICS

The picture below shows a particular organism. Use the picture to answer the questions that follow.

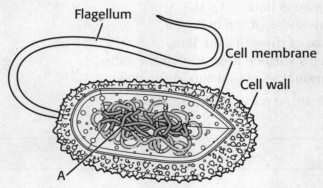

Flagellum

Cell membrane

Cell wall

A

9. What type of organism does the picture represent? How do you know?

_____

_____

10. Which structure helps the organism move?

_____

_____

11. What part of the organism does the letter *A* represent?

_____

_____

Skills Worksheet

Section Review

Eukaryotic Cells

USING KEY TERMS

1. In your own words, write a definition for each of the following terms: *ribosome*, *lysosome*, and *cell wall*.

_____

_____

_____

UNDERSTANDING KEY IDEAS

_____ 2. Which of the following are found mainly in animal cells?
    a. mitochondria
    b. lysosomes
    c. ribosomes
    d. Golgi complexes

3. What is the function of a Golgi complex? What is the function of the endoplasmic reticulum?

_____

_____

_____

_____

CRITICAL THINKING

4. **Making Comparisons** Describe three ways in which plant cells differ from animal cells.

_____

_____

_____

5. **Applying Concepts** Every cell needs ribosomes. Explain why.

_____

_____

| Section Review *continued*

6. Predicting Consequences A certain virus attacks the mitochondria in cells. What would happen to a cell if all of its mitochondria were destroyed?

_____

_____

7. Expressing Opinions Do you think that having chloroplasts gives plant cells an advantage over animal cells? Support your opinion.

_____

_____

_____

_____

INTERPRETING GRAPHICS

Use the diagram below to answer the questions that follow.

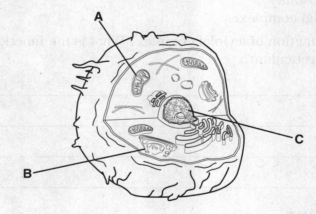

8. Is this a diagram of a plant cell or an animal cell? Explain how you know.

_____

_____

9. What organelle does the letter *B* refer to?

_____

_____

Skills Worksheet

Section Review

The Organization of Living Things

USING KEY TERMS

1. Use each of the following terms in a separate sentence: *tissue*, *organ*, and *function*.

_____

_____

_____

UNDERSTANDING KEY IDEAS

_____ **2.** What are the four levels of organization in living things?
   a. cell, multicellular, organ, organ system
   b. single cell, multicellular, tissue, organ
   c. larger size, longer life, specialized cells, organs
   d. cell, tissue, organ, organ system

MATH SKILLS

3. One multicellular organism is a cube. Each of its sides is 3 cm long. Each of its cells is 1 cm^3. How many cells does it have? If each side doubles in length, how many cells will it then have? Show your work below.

CRITICAL THINKING

4. Applying Concepts Explain the relationship between structure and function. Use alveoli as an example. Be sure to include more than one level of organization.

_____

_____

_____

_____

| Section Review *continued*

5. Making Inferences Why can multicellular organisms be more complex than single-cell organisms? Use the three advantages of being multicellular to help explain your answer.

_____

_____

_____

_____

Skills Worksheet

Chapter Review

USING KEY TERMS

Complete each of the following sentences by choosing the correct term from the word bank.

cell	organ	cell membrane
prokaryote	organelles	eukaryote
cell wall	tissue	structure
function		

1. A(n) _____ is the most basic unit of all living things.

2. The job that an organ does is the _____ of that organ.

3. Ribosomes and mitochondria are types of _____.

4. A(n) _____ is an organism whose cells have a nucleus.

5. A group of cells working together to perform a specific function is

a(n) _____.

6. Only plant cells have a(n) _____.

UNDERSTANDING KEY IDEAS

Multiple Choice

_____ **7.** Which of the following best describes an organ?
   a. a group of cells that work together to perform a specific job
   b. a group of tissues that belong to different systems
   c. a group of tissues that work together to perform a specific job
   d. a body structure, such as muscles or lungs

_____ **8.** The benefits of being multicellular include
   a. small size, long life, and cell specialization.
   b. generalized cells, longer life, and ability to prey on small animals.
   c. larger size, more enemies, and specialized cells.
   d. longer life, larger size, and specialized cells.

_____ **9.** In eukaryotic cells, which organelle contains the DNA?
   a. nucleus   **c.** smooth ER
   b. Golgi complex   **d.** vacuole

_____ **10.** Which of the following statements is part of the cell theory?
   a. All cells suddenly appear by themselves.
   b. All cells come from other cells.
   c. All organisms are multicellular.
   d. All cells have identical parts.

Chapter Review *continued*

_____11. The surface area–to-volume ratio of a cell limits
  a. the number of organelles that the cell has.
  b. the size of the cell.
  c. where the cell lives.
  d. the types of nutrients that a cell needs.

_____12. Two types of organisms whose cells do not have a nucleus are
  a. prokaryotes and eukaryotes.
  b. plants and animals.
  c. bacteria and archaea.
  d. single-celled and multicellular organisms.

Short Answer

13. Explain why most cells are small.

_____

_____

_____

14. Describe the four levels of organization in living things.

_____

_____

_____

15. What is the difference between the structure of an organ and the function of the organ?

_____

_____

_____

16. Name two functions of a cell membrane.

_____

_____

_____

17. What are the structure and function of the cytoskeleton in a cell?

_____

_____

_____

Chapter Review *continued*

CRITICAL THINKING

18. Concept Mapping Use the following terms to create a concept map: *cells, organisms, Golgi complex, organ systems, organs, nucleus, organelle,* and *tissues.*

Holt Science and Technology   **31**   Cells: The Basic Units of Life

19. Making Comparisons Compare and contrast the functions of the endoplasmic reticulum and the Golgi complex.

_____

_____

_____

20. Identifying Relationships Explain how the structure and function of an organism's parts are related. Give an example.

_____

_____

_____

21. Evaluating Hypotheses One of your classmates states a hypothesis that all organisms must have organ systems. Is your classmate's hypothesis valid? Explain your answer.

_____

_____

_____

22. Predicting Consequences What would happen if all of the ribosomes in your cells disappeared?

_____

_____

_____

23. Expressing Opinions Scientists think that millions of years ago the surface of the Earth was very hot and that the atmosphere contained a lot of methane. In your opinion, which type of organism, a bacterium or an archaeon, is the older form of life? Explain your reasoning.

_____

_____

_____

_____

Chapter Review *continued*

INTERPRETING GRAPHICS

Use the diagram below to answer the questions that follow.

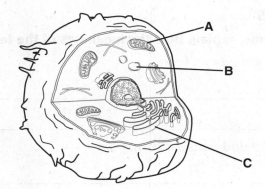

24. What is the name of the structure identified by the letter *A*?

_____

25. Which letter identifies the structure that digests food particles and foreign invaders?

_____

26. Which letter identifies the structure that makes proteins, lipids, and other materials and that contains tubes and passageways that enable substances to move to different places in the cell?

_____

Skills Worksheet

Section Review

Exchange with the Environment

USING KEY TERMS

For each pair of terms, explain how the meanings of the terms differ.

1. *diffusion* and *osmosis*

   _____

   _____

2. *active transport* and *passive transport*

   _____

   _____

3. *endocytosis* and *exocytosis*

   _____

   _____

UNDERSTANDING KEY IDEAS

_____ 4. The movement of particles from a less crowded area to a more
   crowded area requires
   - **a.** sunlight.
   - **b.** energy.
   - **c.** a membrane.
   - **d.** osmosis.

5. What structures allow small particles to cross cell membranes?

   _____

   _____

MATH SKILLS

6. The area of particle 1 is 2.5 mm^2. The area of particle 2 is 0.5 mm^2. The area
   of particle 1 is how many times as big as the area of particle 2? Show your
   work below.

Section Review *continued*

CRITICAL THINKING

7. Predicting Consequences What would happen to a cell if its channel proteins were damaged and unable to transport particles? What would happen to the organism if many of its cells were damaged in this way? Explain your answer.

_____

_____

_____

_____

8. Analyzing Ideas Why does active transport require energy?

_____

_____

_____

Skills Worksheet

Section Review

Cell Energy

USING KEY TERMS

1. In your own words, write a definition for the term *fermentation*.

_____

_____

UNDERSTANDING KEY IDEAS

_____ 2. O_2 is released during
    a. cellular respiration.
    b. photosynthesis.
    c. breathing.
    d. fermentation.

3. How are photosynthesis and cellular respiration related?

_____

_____

4. How are respiration and fermentation similar? How are they different?

_____

_____

MATH SKILLS

5. Cells of plant A make 120 molecules of glucose an hour. Cells of plant B make half as much glucose as plant A does. How much glucose does plant B make every minute? Show your work below.

Section Review *continued*

CRITICAL THINKING

6. Analyzing Relationships Why are plants important to the survival of all other organisms?

_____

_____

_____

_____

_____

7. Applying Concepts You have been given the job of restoring life to a barren island. What types of organisms would you put on the island? If you want to have animals on the island, what other organisms must you bring? Explain your answer.

_____

_____

_____

_____

_____

Skills Worksheet

Section Review

The Cell Cycle

USING KEY TERMS

1. In your own words, write a definition for each of the following terms: *cell cycle* and *cytokinesis*.

_____

_____

UNDERSTANDING KEY IDEAS

_____ 2. Eukaryotic cells
   a. do not divide.
   b. undergo binary fission.
   c. undergo mitosis.
   d. have cell walls.

3. Why is it important for chromosomes to be copied before cell division?

_____

_____

4. Describe mitosis.

_____

_____

MATH SKILLS

5. Cell A takes 6 h to complete division. Cell B takes 8 h to complete division. After 24 h, how many more copies of cell A would there be than cell B? Show your work below.

CRITICAL THINKING

6. Predicting Consequences What would happen if cytokinesis occurred without mitosis?

_____

_____

_____

7. Applying Concepts How does mitosis ensure that a new cell is just like its parent cell?

_____

_____

_____

8. Making Comparisons Compare the processes that animal cells and plant cells use to make new cells. How are the processes different?

_____

_____

_____

_____

Skills Worksheet

Chapter Review

USING KEY TERMS

1. Use the following terms in the same sentence: *diffusion* and *osmosis*.

2. In your own words, write a definition for each of the following terms: *exocytosis* and *endocytosis*.

Complete each of the following sentences by choosing the correct term from the word bank.

> cellular respiration   photosynthesis   fermentation

3. Plants use _____ to make glucose.

4. During _____, oxygen is used to break down food molecules releasing large amounts of energy.

For each pair of terms, explain how the meanings of the terms differ.

5. *cytokinesis* and *mitosis*

_____

_____

_____

6. *active transport* and *passive transport*

_____

_____

_____

7. *cellular respiration* and *fermentation*

_____

_____

_____

Chapter Review *continued*

UNDERSTANDING KEY IDEAS
Multiple Choice

_____ **8.** The process in which particles move through a membrane from a region of low concentration to a region of high concentration is
    a. diffusion.   **c.** active transport.
    b. passive transport.   **d.** fermentation.

_____ **9.** What is the result of mitosis and cytokinesis?
    a. two identical cells   **c.** chloroplasts
    b. two nuclei   **d.** two different cells

_____ **10.** Before the energy in food can be used by a cell, the energy must first be transferred to molecules of
    a. proteins.   **c.** DNA.
    b. carbohydrates.   **d.** ATP.

_____ **11.** Which of the following cells would form a cell plate during the cell cycle?
    a. a human cell   **c.** a plant cell
    b. a prokaryotic cell   **d.** All of the above

Short Answer

12. Are exocytosis and endocytosis examples of active or passive transport? Explain your answer.

_____

_____

_____

13. Name the cell structures that are needed for photosynthesis and the cell structures that are needed for cellular respiration.

_____

_____

_____

14. Describe the three stages of the cell cycle of a eukaryotic cell.

_____

_____

_____

Chapter Review *continued*

CRITICAL THINKING

15. Concept Mapping Use the following terms to create a concept map:
*chromosome duplication, cytokinesis, prokaryote, mitosis, cell cycle,
binary fission,* and *eukaryote.*

Chapter Review *continued*

16. Making Inferences Which one of the plants pictured below was given water mixed with salt, and which one was given pure water? Explain how you know, and be sure to use the word *osmosis* in your answer.

_____

_____

_____

_____

17. Identifying Relationships Why would your muscle cells need to be supplied with more food when there is a lack of oxygen than when there is plenty of oxygen present?

_____

_____

_____

_____

18. Applying Concepts A parent cell has 10 chromosomes.

a. Will the cell go through binary fission or mitosis and cytokinesis to produce new cells?

_____

b. How many chromosomes will each new cell have after the parent cell divides?

_____

Chapter Review *continued*

INTERPRETING GRAPHICS

The picture below shows a cell. Use the picture below to answer the questions that follow.

19. Is the cell prokaryotic or eukaryotic?

_____

_____

_____

20. Which stage of the cell cycle is this cell in?

_____

_____

_____

21. How many chromatids are present? How many pairs of homologous chromosomes are present?

_____

_____

_____

22. How many chromosomes will be present in each of the new cells after the cell divides?

_____

_____

_____

Skills Worksheet

Section Review

Mendel and His Peas
USING KEY TERMS

1. Use each of the following terms in a separate sentence: *heredity*, *dominant trait*, and *recessive trait*.

_____

_____

_____

UNDERSTANDING KEY IDEAS

_____ 2. A plant that has both male and female reproductive structures is able to
   a. self-replicate.
   b. self-pollinate.
   c. change colors.
   d. None of the above

3. Explain the difference between self-pollination and cross-pollination.

_____

_____

_____

4. What is the difference between a trait and a characteristic? Give one example of each.

_____

_____

5. Describe Mendel's first set of experiments.

_____

_____

6. Describe Mendel's second set of experiments.

_____

_____

MATH SKILLS

7. In a bag of chocolate candies there are 21 brown candies and 6 green candies. What is the ratio of brown to green? What is the ratio of green to brown? Show your work below.

CRITICAL THINKING

8. Predicting Consequences Gregor Mendel used only true-breeding plants. If he had used plants that were not true breeding, do you think he would have discovered dominant and recessive traits? Explain.

_____

_____

_____

9. Applying Concepts In cats, there are two types of ears: normal and curly. A curly eared cat mated with a normal eared cat and all of the kittens had curly ears. Are curly ears a dominant or recessive trait? Explain.

_____

_____

_____

10. Identifying Relationships List three other fields of study that use ratios.

_____

_____

_____

Skills Worksheet

Section Review

Traits and Inheritance
USING KEY TERMS

1. Use the following terms in the same sentence: *gene* and *allele*.

_____

_____

2. In your own words, write a definition for each of the following terms: *genotype* and *phenotype*.

_____

_____

UNDERSTANDING KEY IDEAS

_____ 3. Use a Punnett square to determine the possible genotypes of the offspring of a *BB* x *Bb* cross.
   a. all *BB*
   b. *BB, Bb*
   c. *BB, Bb, bb*
   d. all *bb*

4. How are genes and alleles related to genotype and phenotype?

_____

_____

5. Describe three exceptions to Mendel's observations.

_____

_____

MATH SKILLS

6. What is the probability of rolling a five on one die three times in a row? Show your work below.

CRITICAL THINKING

7. Applying Concepts The allele for a cleft chin, *C*, is dominant among humans. What are the results of a cross between parents with genotypes *Cc* and *cc*?

_____

_____

_____

_____

INTERPRETING GRAPHICS

The Punnett Square below shows the alleles for full color in rabbits. Black fur, *B*, is dominant over white fur, *b*.

	?	?
?	*Bb*	*Bb*
?	*Bb*	*Bb*

8. Given the combinations shown, what are the genotypes of the parents?

_____

_____

9. If black fur had incomplete dominance over white fur, what color would the offspring be?

_____

_____

Skills Worksheet

Section Review

Meiosis

USING KEY TERMS

1. Use each of the following terms in the same sentence: *meiosis* and *sex chromosomes*.

_____

_____

In each of the following sentences, replace the incorrect term with the correct term from the word bank.

pedigree   homologous chromosomes
meiosis   mitosis

2. During fertilization, chromosomes are copied, and then the nucleus divides twice.

_____

_____

3. A Punnett square is used to show how inherited traits move through a family.

_____

_____

4. During meiosis, sex cells line up in the middle of the cell.

_____

_____

UNDERSTANDING KEY IDEAS

_____ **5.** Genes are found on
   a. chromosomes.
   b. proteins.
   c. alleles.
   d. sex cells.

6. If there are 14 chromosomes in pea plant cells, how many chromosomes are present in a sex cell of a pea?

_____

_____

Name _____ Class _____ Date _____

7. Draw the eight steps of meiosis. Label one chromosome, and show its position in each step.

INTERPRETING GRAPHICS

Use this pedigree to answer the question below.

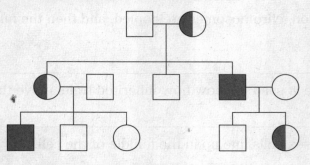

8. Is this disorder sex linked? Explain your reasoning.

_____

_____

CRITICAL THINKING

9. Identifying Relationships Put the following in order of smallest to largest: chromosome, gene, and cell.

_____

_____

10. Applying Concepts A pea plant has purple flowers. What alleles for flower color could the sex cells carry?

_____

_____

Name _____ Class _____ Date _____

Chapter Review

USING KEY TERMS

Complete each of the following sentences by choosing the correct term from the word bank.

sex cells	genotype	sex chromosomes
alleles	phenotype	meiosis

1. Sperm and eggs are known as _____.

2. The _____ is the expression of a trait and is determined by the combination of alleles called the _____.

3. _____ produces cells with half the normal number of chromosomes.

4. Different versions of the same genes are called _____.

UNDERSTANDING KEY IDEAS

Multiple Choice

_____ 5. Genes carry information that determines
   a. alleles.
   b. ribosomes.
   c. chromosomes.
   d. traits.

_____ 6. The process that produces sex cells is
   a. mitosis.
   b. photosynthesis.
   c. meiosis.
   d. probability.

_____ 7. The passing of traits from parents to offspring is called
   a. probability.
   b. heredity.
   c. recessive.
   d. meiosis.

_____ 8. If you cross a white flower with the genotype *pp* with a purple flower with the genotype *PP*, the possible genotypes in the offspring are
   a. *PP* and *pp*.
   b. all *Pp*.
   c. all *PP*.
   d. all *pp*.

Chapter Review *continued*

_____ 9. For the cross in item 8, what would the phenotypes be?
   a. all white
   b. 3 purple and 1 white
   c. all purple
   d. half white, half purple

_____ 10. In meiosis,
   a. chromosomes are copied twice.
   b. the nucleus divides once.
   c. four cells are produced from a single cell.
   d. two cells are produced from a single cell.

_____ 11. When one trait is not completely dominant over another, it is called
   a. recessive.
   b. incomplete dominance.
   c. environmental factors.
   d. uncertain dominance.

Short Answer

12. Which sex chromosomes do females have? Which do males have?

_____

_____

_____

_____

13. In one or two sentences, define the term *recessive trait* in your own words.

_____

_____

_____

_____

14. How are sex cells different from other body cells?

_____

_____

_____

_____

_____

▌Chapter Review *continued*

15. What is a sex-linked disorder? Give one example of a sex-linked disorder that is found in humans.

_____

_____

_____

_____

Critical Thinking

16. Concept Mapping Use the following terms to create a concept map: *meiosis, eggs, cell division, X chromosome, mitosis, Y chromosome, sperm,* and *sex cells.*

Chapter Review *continued*

17. Identifying Relationships If you were a carrier of one allele for a certain recessive disorder, how could genetic counseling help you prepare for the future?

_____

_____

_____

_____

_____

18. Applying Concepts If a child has blond hair and both of her parents have brown hair, what does that tell you about the allele for blond hair? Explain.

_____

_____

_____

_____

_____

19. Applying Concepts What is the genotype of a pea plant that is true-breeding for purple flowers?

_____

_____

_____

_____

Chapter Review *continued*

INTERPRETING GRAPHICS

Use the Punnett square below to answer the questions that follow.

	?	?
T	TT	TT
t	Tt	Tt

20. What is the unknown genotype?

_____

_____

_____

21. If *T* represents the allele for tall pea plants and *t* represents the allele for short pea plants, what is the phenotype of each parent and of the offspring?

_____

_____

22. If each of the offspring were allowed to self-fertilize, what are the possible genotypes in the next generation?

_____

_____

_____

23. What is the probability of each genotype in item 22?

_____

_____

Skills Worksheet

Section Review

What Does DNA Look Like?

USING KEY TERMS

1. Use the term *DNA* in a sentence.

   _____

2. In your own words, write a definition for the term *nucleotide*.

   _____

UNDERSTANDING KEY IDEAS

3. List three important events that led to understanding the structure of DNA.

   _____

   _____

   _____

_____ 4. Which of the following is NOT part of a nucleotide?
   a. base
   b. sugar
   c. fat
   d. phosphate

MATH SKILLS

5. If a sample of DNA contained 20% cytosine, what percentage of guanine would be in this sample? What percentage of adenine would be in the sample? Explain. Show your work below.

Name _____ Class _____ Date _____

CRITICAL THINKING

6. Making Inferences Explain what is meant by the statement "DNA unites all organisms."

_____

_____

_____

7. Applying Concepts What would the complementary strand of DNA be for the sequence of bases below?

C T T A G G C T T A C C A

_____

_____

_____

8. Analyzing Processes How are copies of DNA made? Draw a picture as part of your answer.

_____

_____

_____

_____

Skills Worksheet

Section Review

How DNA Works

USING KEY TERMS

1. Use each of the following terms in the same sentence: *ribosome* and *RNA*.

_____

_____

2. In your own words, write a definition for the term *mutation*.

_____

_____

UNDERSTANDING KEY IDEAS

3. Explain the relationship between genes and proteins.

_____

_____

4. List three possible types of mutations.

_____

_____

_____

_____ **5.** Which type of mutation causes sickle cell anemia?
　　　　　a. substitution　　　　　　**c.** deletion
　　　　　b. insertion　　　　　　　 **d.** mutagen

MATH SKILLS

6. A set of 23 chromosomes in a human cell contains 3.2 billion pairs of DNA bases in sequence. How many pairs of bases are in each chromosome? Show your work below.

| Section Review *continued*

CRITICAL THINKING

7. Applying Concepts In which cell type might a mutation be passed from generation to generation? Explain.

_____

_____

8. Making Comparisons How is genetic engineering different from natural reproduction?

_____

_____

INTERPRETING GRAPHICS

The illustration below shows a sequence of bases on one strand of a DNA molecule. Use the illustration below to answer the questions that follow.

a.

A C T C C T G A A

b.

9. How many amino acids are coded for by the sequence on one side (A) of this DNA strand?

_____

_____

10. What is the order of bases on the complementary side of the strand (B), from left to right?

_____

_____

11. If a G were inserted as the first base on the top side (A), what would the order of bases be on the complementary side (B)?

_____

_____

Skills Worksheet)

Chapter Review

USING KEY TERMS

1. Use the following terms in the same sentence: *mutation* and *mutagen*.

_____

_____

The statements below are false. For each statement, replace the underlined term to make a true statement.

_____ **2.** The information in DNA is coded in the order of <u>amino acids</u> along one side of the DNA molecule.

_____ **3.** The "factory" that assembles proteins based on the DNA code is called a <u>gene</u>.

UNDERSTANDING KEY IDEAS

Multiple Choice

_____ **4.** James Watson and Francis Crick
   a. took X-ray pictures of DNA.
   b. discovered that genes are in chromosomes.
   c. bred pea plants to study heredity.
   d. made models to figure out DNA's shape.

_____ **5.** In a DNA molecule, which of the following bases pair together?
   a. adenine and cytosine
   b. thymine and adenine
   c. thymine and guanine
   d. cytosine and thymine

_____ **6.** A gene can be all of the following EXCEPT
   a. a set of instructions for a trait.
   b. a complete chromosome.
   c. instructions for making a protein.
   d. a portion of a strand of DNA.

_____ **7.** Which of the following statements about DNA is NOT true?
   a. DNA is found in all organisms.
   b. DNA is made up of five subunits.
   c. DNA has a structure like a twisted ladder.
   d. Mistakes can be made when DNA is copied.

Chapter Review *continued*

_____ **8.** Within the cell, where are proteins assembled?
     a. the cytoplasm
     b. the nucleus
     c. the amino acids
     d. the chromosomes

_____ **9.** Changes in the type or order of the bases in DNA are called
     a. nucleotides.
     b. mutations.
     c. RNA.
     d. genes.

Short Answer

10. What would be the complementary strand of DNA for the following sequence of bases?

C T T A G G C T T A C C A

_____

11. If the DNA sequence TGAGCCATGA is changed to TGAGCACATGA, what kind of mutation has occurred?

_____

_____

12. Explain how the DNA in genes relates to the traits of an organism.

_____

_____

13. Why is DNA frequently found associated with proteins inside of cells?

_____

_____

14. What is the difference between DNA and RNA?

_____

_____

CRITICAL THINKING

15. Concept Mapping Use the following terms to create a concept map: *bases*, *adenine*, *thymine*, *nucleotides*, *guanine*, *DNA*, and *cytosine*.

16. Analyzing Processes Draw and label a picture that explains how DNA is copied.

17. Analyzing Processes Draw and label a picture that explains how proteins are made.

| Chapter Review *continued*

18. Applying Concepts The following DNA sequence codes for how many amino acids?

T C A G C C A C C T A T G G A

_____

19. Making Inferences Why does the government make laws about the use of chemicals that are known to be mutagens?

_____

_____

INTERPRETING GRAPHICS

The illustration below shows the process of replication of a DNA strand. Use this illustration to answer the questions that follow.

_____**20.** Which strands are part of the original molecule?
  a. A and B
  b. A and C
  c. A and D
  d. None of the above

_____**21.** Which strands are new?
  a. A and B
  b. B and C
  c. C and D
  d. None of the above

_____**22.** Which strands are complementary?
  a. A and C
  b. B and C
  c. All of the strands
  d. None of the strands

Section Review

Change over Time
USING KEY TERMS

Complete each of the following sentences by choosing the correct term from the word bank.

adaptation	species
fossil	evolution

1. Members of the same _____ can mate with one another to produce offspring.

2. A(n) _____ helps an organism survive.

3. When populations change over time, _____ has occurred.

UNDERSTANDING KEY IDEAS

_____ **4.** A human's arm, a cat's front leg, a dolphin's front flipper, and a bat's wing
    a. have similar kinds of bones.
    b. are used in similar ways.
    c. are very similar to insect wings and jellyfish tentacles.
    d. have nothing in common.

5. How does the fossil record show that species have changed over time?

_____

_____

_____

_____

6. What evidence do fossils provide about the ancestors of whales?

_____

_____

_____

_____

Section Review *continued*

CRITICAL THINKING

7. Making Comparisons Other than the examples provided in the text, how are whales different from fishes?

_____

_____

_____

_____

8. Forming Hypotheses Is a person's DNA likely to be more similar to the DNA of his or her biological parents or to the DNA of one of his or her cousins? Explain your answer.

_____

_____

_____

_____

INTERPRETING GRAPHICS

9. The photograph in your textbook for this Section Review shows the layers of sedimentary rock exposed during the construction of a road. Imagine that a species that lived 200 million years ago is found in layer **b**. Would the species' ancestor, which lived 250 million years ago, most likely be found in layer **a** or in layer **c**? Explain your answer.

_____

_____

_____

_____

Skills Worksheet

Section Review

How Does Evolution Happen?
USING KEY TERMS

1. In your own words, write a definition for the term *trait*.

_____

_____

2. Use the following terms in the same sentence: *selective breeding* and *natural selection*.

_____

_____

UNDERSTANDING KEY IDEAS

_____ **3.** Modern scientific explanations of evolution
  a. have replaced Darwin's theory.
  b. rely on genetics instead of natural selection.
  c. fail to explain how traits are inherited.
  d. combine the principles of natural selection and genetic inheritance.

4. Describe the observations that Darwin made about the species on the Galápagos Islands.

_____

_____

5. Summarize the ideas that Darwin developed from books by Malthus and Lyell.

_____

_____

_____

6. Describe the four parts of Darwin's theory of evolution by natural selection.

_____

_____

_____

_____

| Section Review *continued*

7. What knowledge did Darwin lack that modern scientists now use to explain evolution?

_____

_____

_____

_____

MATH SKILLS

8. In a sample of 80 beetles, 50 beetles had 4 spots each, and the rest had 6 spots each. What was the average number of spots per beetle? Show your work below.

CRITICAL THINKING

9. Making Comparisons In selective breeding, humans influence the course of evolution. What determines the course of evolution in natural selection?

_____

_____

_____

10. Predicting Consequences Suppose an island in the Pacific Ocean was just formed by a volcano. Over the next million years, how might species evolve on this island?

_____

_____

_____

_____

Skills Worksheet

Section Review

Natural Selection in Action
USING KEY TERMS

1. In your own words, write a definition for the term *speciation*.

   _____

   _____

UNDERSTANDING KEY IDEAS

_____ 2. Two populations have evolved into two species when
      a. the populations are separated.
      b. the populations look different.
      c. the populations can no longer interbreed.
      d. the populations adapt.

3. Explain why the number of tuskless elephants in Uganda may be increasing.

   _____

   _____

   _____

MATH SKILLS

4. A female cockroach can produce 80 offspring at a time. If half of the offspring produced by a certain female are female and each female produces 80 offspring, how many cockroaches are there in the third generation? Show your work below.

Name _____ Class _____ Date _____

CRITICAL THINKING

5. Forming Hypotheses Most kinds of cactus have leaves that grow in the form of spines. The stems or trunks become thick, juicy pads or barrels. Explain how these cactus parts might have evolved.

_____

_____

_____

_____

_____

_____

6. Making Comparisons Suggest an organism other than an insect that might evolve an adaptation to human activities.

_____

_____

_____

_____

_____

_____

Skills Worksheet)

Chapter Review

USING KEY TERMS

Complete each of the following sentences by choosing the correct term from the word bank.

adaptation   evolution   species
natural selection   generation time   speciation
fossil record   selective breeding

1. When a single population evolves into two populations that cannot interbreed anymore, _____ has occurred.

2. Darwin's theory of _____ explained the process by which organisms become well-adapted to their environment.

3. A group of organisms that can mate with each other to produce offspring is known as a(n) _____.

4. The _____ provides information about organisms that have lived in the past.

5. In _____, humans select organisms with desirable traits that will be passed from one generation to another.

6. A(n) _____ helps an organism survive better in its environment.

7. Populations of insects and bacteria can evolve quickly because they usually have a short _____.

UNDERSTANDING KEY IDEAS
Multiple Choice

_____ 8. Fossils are commonly found in
   a. sedimentary rock.   c. granite.
   b. all kinds of rock.   d. loose sand.

_____ 9. The fact that all organisms have DNA as their genetic material is evidence that
   a. all organisms undergo natural selection.
   b. all organisms may have descended from a common ancestor.
   c. selective breeding takes place every day.
   d. genetic resistance rarely occurs.

_____**10.** Charles Darwin puzzled over differences in the _____ of the different species of Galápagos finches.

    a. webbed feet   **c.** bone structure of the wings

    b. beaks   **d.** eye color

_____**11.** Darwin observed variations among individuals within a population, but he did not realize that these variations were caused by

    a. interbreeding.   **c.** differences in genes.

    b. differences in food.   **d.** selective breeding.

Short Answer

12. Identify two ways that organisms can be compared to provide evidence of evolution from a common ancestor.

_____

_____

_____

13. Describe evidence that supports the hypothesis that whales evolved from land-dwelling mammals.

_____

_____

_____

14. Why are some animals more likely to survive to adulthood than other animals are?

_____

_____

_____

15. Explain how genetics is related to evolution.

_____

_____

_____

16. Outline an example of the process of speciation.

_____

_____

_____

_____

_____

Chapter Review *continued*

CRITICAL THINKING

17. Concept Mapping Use the following terms to create a concept map:
struggle to survive, theory, genetic variation, Darwin, overpopulation,
natural selection, and *successful reproduction.*

| Chapter Review *continued*

18. Making Inferences How could natural selection affect the songs that birds sing?

_____

_____

_____

_____

19. Forming Hypotheses In Australia, many animals look like mammals from other parts of the world. But most of the mammals in Australia are marsupials, which carry their young in pouches after birth. Few kinds of marsupials are found anywhere else in the world. What is a possible explanation for the presence of so many of these unique mammals in Australia?

_____

_____

_____

_____

20. Analyzing Relationships Geologists have evidence that the continents were once a single giant continent. This giant landform eventually split apart, and the individual continents moved to their current positions. What role might this drifting of continents have played in evolution?

_____

_____

_____

_____

| Chapter Review *continued*

INTERPRETING GRAPHICS

The graphs below show information about the infants that are born and the infants that have died in a population. The weight of each infant was measured at birth. Use the graphs to answer the questions that follow.

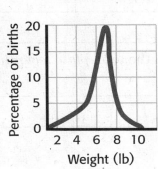

Infant Births by Birth Weight

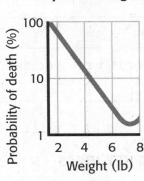

Infant Deaths by Birth Weight

21. What is the most common birth weight?

_____

_____

22. At which birth weight is an infant most likely to survive?

_____

_____

23. How do the principles of natural selection help explain why there are more deaths among babies whose birth weights are low than among babies whose birth weights are average?

_____

_____

_____

Skills Worksheet

Section Review

Evidence of the Past
USING KEY TERMS

1. Use the following terms in the same sentence: *fossil* and *extinct*.

   _____

   _____

2. In your own words, write a definition for the term *plate tectonics*.

   _____

   _____

UNDERSTANDING KEY IDEAS

3. Explain how a fossil forms in sedimentary rock.

   _____

   _____

4. What kind of information does the geologic time scale show?

   _____

   _____

5. About how many years of Earth's history was Precambrian time?

   _____

   _____

6. What are two possible causes of mass extinctions?

   _____

   _____

| Section Review *continued*

MATH SKILLS

7. The Earth formed 4.6 billion years ago. Modern humans have existed for about 160,000 years. Simple worms have existed for at least 500 million years. For what fraction of the history of Earth have humans existed? have worms existed? Show your work below.

CRITICAL THINKING

8. Identifying Relationships Why are both absolute dating and relative dating used to determine the age of fossils?

_____

_____

_____

_____

9. Making Inferences Fossils of *Mesosaurus*, the small aquatic reptile shown in your textbook, have been found only in Africa and South America. Using what you know about plate tectonics, how would you explain this finding?

_____

_____

_____

_____

Section Review

Eras of the Geologic Time Scale
USING KEY TERMS

1. Use each of the following terms in a separate sentence: *Precambrian time*, *Paleozoic era*, *Mesozoic era*, and *Cenozoic era*.

_____

_____

_____

_____

UNDERSTANDING KEY IDEAS

_____ 2. Unlike the atmosphere today, the atmosphere 3.5 billion years ago did not contain
   a. carbon dioxide.
   b. nitrogen.
   c. gases.
   d. ozone.

3. How do prokaryotic cells and eukaryotic cells differ?

_____

_____

4. Explain why cyanobacteria were important to the development of life on Earth.

_____

_____

5. Place in chronological order the following events on Earth:
   a. The first cells appeared that could make their own food from sunlight.
   b. The ozone layer formed.
   c. Simple chemicals reacted to form the molecules of life.
   d. Animals appeared.
   e. The first organisms appeared.
   f. Humans appeared.
   g. The Earth formed.

_____

_____

| Section Review *continued*

MATH SKILLS

6. Calculate the total number of years that each of the geologic eras lasted, rounding to the nearest 100 million. Then, calculate each of these values as a percentage of the total 4.6 billion years of Earth's history. Round your answer to the units place. Show your work below.

CRITICAL THINKING

7. Making Inferences Which chemicals probably made up the first cells on Earth?

_____

_____

_____

_____

_____

8. Forming Hypotheses Think of your own hypothesis to explain the disappearance of the dinosaurs. Explain your hypothesis.

_____

_____

_____

_____

_____

Skills Worksheet

Section Review

Humans and Other Primates
USING KEY TERMS

1. Use each of the following words in the same sentence: *primate*, *hominid*, and *Homo sapiens*.

_____

_____

_____

UNDERSTANDING KEY IDEAS

_____ 2. The unique characteristics of primates are
   a. bipedalism and thumbs.
   b. opposable thumbs.
   c. opposable thumbs and binocular vision.
   d. opposable toes and thumbs.

3. Describe the major evolutionary developments from early hominids to modern humans.

_____

_____

4. Compare members of the *Homo* group with australopithecines.

_____

_____

CRITICAL THINKING

5. **Forming Hypotheses** Suggest some reasons why Neanderthals might have become extinct.

_____

_____

_____

_____

Section Review *continued*

6. Making Inferences Imagine you are a scientist excavating an ancient camp-site. What might you infer about the people who used the site if you found the charred bones of large animals and various stone blades among human fossils?

_____

_____

_____

_____

_____

_____

INTERPRETING GRAPHICS

The figure below shows a possible ancestral relationship between humans and some modern apes. Use this figure to answer the questions that follow.

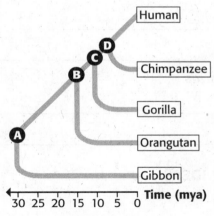

7. Which letter represents the ancestor of all the apes?

_____

8. To which living ape are gorillas most closely related?

_____

_____

_____

Name _____ Class _____ Date _____

Chapter Review

USING KEY TERMS

Complete each of the following sentences by choosing the correct term from the word bank.

Precambrian time	Paleozoic era
Mesozoic era	Cenozoic era

1. During _____, life is thought to have originated from

nonliving matter.

2. The Age of Mammals refers to the _____.

3. The Age of Reptiles refers to the _____.

4. Plants colonized land during the _____.

For each pair of terms, explain how the meanings of the terms differ.

5. *relative dating* and *absolute dating*

_____

_____

6. *primates* and *hominids*

_____

_____

UNDERSTANDING KEY IDEAS

Multiple Choice

_____ **7.** If the half-life of an unstable element is 5,000 years, what percentage
of the parent material will be left after 10,000 years?
  a. 100%
  b. 75%
  c. 50%
  d. 25%

_____ **8.** The first cells on Earth appeared in
  a. Precambrian time.
  b. the Paleozoic era.
  c. the Mesozoic era.
  d. the Cenozoic era.

_____ **9.** In which era are we currently living?
  a. Precambrian time
  b. Paleozoic era
  c. Mesozoic era
  d. Cenozoic era

_____ **10.** Scientists think that the closest living relatives of humans are
  a. lemurs.
  b. monkeys.
  c. gorillas.
  d. chimpanzees.

Short Answer

11. Describe how plant and animal remains can become fossils.

_____

_____

12. What information do fossils provide about the history of life?

_____

_____

13. List three important steps in the early development of life on Earth.

_____

_____

_____

14. List two important groups of organisms that appeared during each of the three most recent geologic eras.

_____

_____

15. Describe the event that scientists think caused the mass extinction at the end of the Mesozoic era.

_____

_____

16. From which geologic era are fossils most commonly found?

_____

_____

17. Describe two characteristics that are shared by all primates.

_____

_____

18. Which hominid species is alive today?

_____

_____

CRITICAL THINKING

19. **Concept Mapping** Use the following terms to create a concept map: *Earth's history, humans, Paleozoic era, dinosaurs, Precambrian time, land plants, Mesozoic era, cyanobacteria,* and *Cenozoic era.*

Chapter Review *continued*

20. Applying Concepts Can footprints be fossils? Explain your answer.

_____

_____

_____

21. Making Inferences If you find rock layers containing fish fossils in a desert, what can you infer about the history of the desert?

_____

_____

_____

22. Applying Concepts Explain how an environmental change can threaten the survival of a species. Give two examples.

_____

_____

_____

23. Analyzing Ideas Why do scientists think the first cells did not need oxygen to survive?

_____

_____

_____

24. Identifying Relationships How does the extinction that occurred at the end of the Mesozoic era relate to the Age of Mammals?

_____

_____

_____

25. Making Comparisons Make a table listing the similarities and differences between australopithecines, early members of the group *Homo*, and modern members of the species *Homo sapiens*.

Chapter Review *continued*

INTERPRETING GRAPHICS

The graph below shows data about fossilized teeth that were found within a series of rock layers. Use this graph to answer the questions that follow.

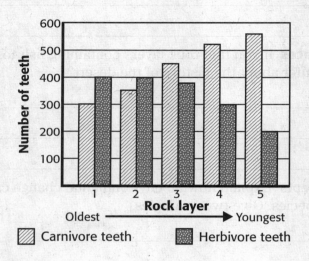

_____26. Which of the following statements best describes the information presented in the graph?
  a. Over time, the number of carnivores decreased and the number of herbivores increased.
  b. Over time, the number of carnivores increased and the number of herbivores increased.
  c. Over time, the number of carnivores and herbivores remained the same.
  d. Over time, the number of carnivores increased and the number of herbivores decreased.

_____27. At what point did carnivore teeth begin to outnumber herbivore teeth?
  a. between layer 1 and layer 2
  b. between layer 2 and layer 3
  c. between layer 3 and layer 4
  d. between layer 4 and layer 5

Skills Worksheet

Section Review

Sorting It All Out
USING KEY TERMS

1. In your own words, write a definition for each of the following terms: *classification* and *taxonomy*.

_____

_____

UNDERSTANDING KEY IDEAS

_____ 2. The two parts of a scientific name are the names of the genus and the
   a. specific name.
   b. phylum name.
   c. family name.
   d. order name.

3. Why do scientists use scientific names for organisms?

_____

_____

4. List the eight levels of classification.

_____

_____

_____

_____

_____

_____

5. Describe how a dichotomous key helps scientists identify organisms.

_____

_____

| Section Review *continued*

CRITICAL THINKING

6. Analyzing Processes Biologists think that millions of species are not classified yet. Why do you think so many species have not been classified yet?

_____

_____

_____

_____

7. Applying Concepts Both dolphins and sharks have a tail and fins. How can you determine if dolphins and sharks are closely related?

_____

_____

_____

_____

INTERPRETING GRAPHICS

Use the figure below to answer the questions that follow.

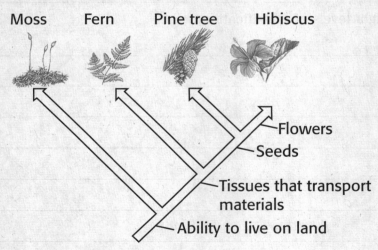

8. Which plant is most similar to the hibiscus?

_____

_____

9. Which plant is least similar to the hibiscus?

_____

_____

Skills Worksheet

Section Review

Domains and Kingdoms
USING KEY TERMS
For each pair of terms, explain how the meanings of the terms differ.

1. *Archaea* and *Bacteria*

_____

_____

2. *Plantae* and *Fungi*

_____

_____

UNDERSTANDING KEY IDEAS

_____ **3.** Biological classification schemes change
   a. as new evidence and more kinds of organisms are discovered.
   b. every 100 years.
   c. when scientists disagree.
   d. only once.

4. Describe the characteristics of each of the three domains.

_____

_____

_____

5. Describe the four kingdoms of domain Eukarya.

_____

_____

_____

_____

MATH SKILLS

6. A certain bacterium can divide every 30 min. If you begin with 1 bacterium, when will you have more than 1,000 bacteria? Show your work below.

CRITICAL THINKING

7. Identifying Relationships How are bacteria similar to fungi? How are fungi similar to animals?

_____

_____

_____

8. Analyzing Methods Why do you think Linnaeus did not include classification kingdoms for categories of archaea and bacteria?

_____

_____

_____

9. Applying Concepts The Venus' flytrap does not move around. It can make its own food by using photosynthesis. It can also trap insects and digest the insects to get nutrients. The flytrap also has a cell wall. Into which kingdom would you place the Venus' flytrap? What makes this organism unusual in the kingdom you chose?

_____

_____

_____

_____

Skills Worksheet

Chapter Review

USING KEY TERMS

Complete each of the following sentences by choosing the correct term from the word bank.

Animalia	Protista	Bacteria
Plantae	Archaea	classification
taxonomy		

1. Linnaeus founded the science of _____.

2. Prokaryotes that live in extreme environments are in the domain

_____.

3. Complex multicellular organisms that can usually move around and respond

to their environment are in the kingdom _____.

4. A system of _____ can help group animals into categories.

5. Prokaryotes that can cause diseases are in the domain

_____.

UNDERSTANDING KEY IDEAS

Multiple Choice

_____ **6.** Scientists classify organisms by
   a. arranging the organisms in orderly groups.
   b. giving the organisms many common names.
   c. deciding whether the organisms are useful.
   d. using only existing categories of classification.

_____ **7.** When the eight levels of classification are listed from broadest to
   narrowest, which level is sixth in the list?
   a. class
   b. order
   c. genus
   d. family

_____ **8.** The scientific name for the European white waterlily is *Nymphaea
   alba*. To which genus does this plant belong?
   a. *Nymphaea*
   b. *alba*
   c. water lily
   d. alba lily

_____ 9. *Animalia*, *Protista*, *Fungi*, and *Plantae* are the
   a. scientific names of different organisms.
   b. names of kingdoms.
   c. levels of classification.
   d. scientists who organized taxonomy.

_____ 10. The simple, single-celled organisms that live in your intestines are classified in the kingdom
   a. Protista.
   b. Bacteria.
   c. Archaea.
   d. Eukarya.

_____ 11. What kind of organism thrives in hot springs and other extreme environments?
   a. fungus
   b. bacterium
   c. archaean
   d. protist

Short Answer

12. Why is the use of scientific names important in biology?

_____

_____

13. What kind of evidence is used by modern taxonomists to classify organisms based on evolutionary relationships?

_____

_____

14. Is a bacterium a type of eukaryote? Explain your answer.

_____

_____

_____

15. Scientists used to classify organisms as either plants or animals. Why doesn't that classification system work?

_____

_____

_____

CRITICAL THINKING

16. Concept Mapping Use the following terms to create a concept map: *kingdom, fern, lizard, Animalia, Fungi, algae, Protista, Plantae,* and *mushroom.*

17. Analyzing Methods Explain how the levels of classification depend on the similarities and differences between organisms.

_____

_____

18. Making Inferences Explain why two species that belong to the same genus, such as white oak (*Quercus alba*) and cork oak (*Quercus suber*), also belong to the same family.

_____

_____

19. Identifying Relationships What characteristic do the members of the four kingdoms of the domain Eukarya have in common?

_____

_____

Chapter Review *continued*

INTERPRETING GRAPHICS

Use the branching diagram of selected primates below to answer the questions that follow.

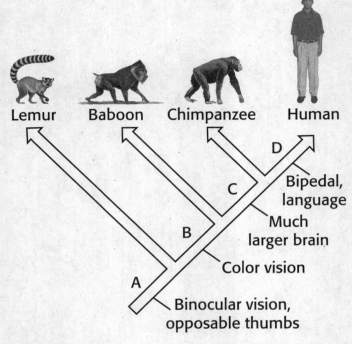

20. Which primate is the closest relative to the common ancestor of all primates?

_____

_____

21. Which primate shares the most traits with humans?

_____

_____

22. Do both lemurs and humans have the characteristics listed at point D? Explain your answer.

_____

_____

23. What characteristic do baboons have that lemurs do not have? Explain your answer.

_____

_____

Skills Worksheet)

Section Review

Bacteria and Archaea
USING KEY TERMS

The statements below are false. For each statement, replace the underlined term to make a true statement.

1. Bacteria are <u>eukaryotes.</u>

_____

_____

2. Bacteria reproduce by <u>primary fission.</u>

_____

_____

UNDERSTANDING KEY IDEAS

_____ **3.** The structure that helps some bacteria survive harsh conditions is called a(n)

    a. endospore.   **c.** exospore.

    b. shell.   **d.** exoskeleton.

4. How are bacteria and archaea different?

_____

_____

5. Draw and label the four stages of binary fission.

❙ Section Review *continued*

6. Describe one advantage of each shape of bacteria.

_____

_____

7. What two things do producer bacteria and plants have in common?

_____

_____

MATH SKILLS

8. An ounce (oz) is equal to about 28 g. If 1 g of soil contains 2.5 billion bacteria, how many bacteria are in 1 oz of soil? Show your work below.

CRITICAL THINKING

9. Applying Concepts Many bacteria cannot reproduce in cooler temperatures and are destroyed at high temperatures. How do humans take advantage of this fact when preparing and storing food?

_____

_____

_____

10. Making Comparisons Scientists are studying cold and dry environments on Earth that are like the environment on Mars. What kind of prokaryotes do you think they might find in these environments on Earth? Explain.

_____

_____

_____

11. Forming Hypotheses You are studying a lake and the prokaryotes that live in it. What conditions of the lake would you measure to form a hypothesis about the kind of prokaryotes that live on the lake?

_____

_____

Skills Worksheet

Section Review

Bacteria's Role in the World

USING KEY TERMS

1. In your own words, write a definition for the term *bioremediation*.

   _____

2. Use the following terms in the same sentence: *pathogenic bacteria* and *antibiotic*.

   _____

   _____

UNDERSTANDING KEY IDEAS

3. What are two ways that bacteria affect plants?

   _____

   _____

4. How can bacteria both cause and cure diseases?

   _____

   _____

5. Explain two ways in which bacteria are crucial to life on Earth.

   _____

   _____

6. Describe two ways in which your life was affected by bacteria today.

   _____

   _____

| Section Review *continued*

MATH SKILLS

7. Nitrogen makes up 78% of air. If you have 2 L of air, how many liters of nitrogen are in the air? Show your work below.

CRITICAL THINKING

8. Identifying Relationships Legumes, which include peas and beans, are efficient nitrogen fixers. Legumes are also a good source of amino acids. What chemical element would you expect to find in amino acids?

_____

_____

_____

_____

9. Applying Concepts Design a bacterium that will be genetically engineered. What do you want it to do? How would it help people or the environment?

_____

_____

_____

_____

Skills Worksheet

Section Review

Viruses

USING KEY TERMS

1. Use the following terms in the same sentence: *virus* and *host*.

_____

_____

UNDERSTANDING KEY IDEAS

_____ **2.** One characteristic viruses have in common with living things is that they

    a. eat.   **c.** sleep.

    b. reproduce.   **d.** grow.

3. Describe the four steps in the lytic cycle.

_____

_____

_____

_____

4. Explain how the lytic cycle and the lysogenic cycle are different.

_____

_____

_____

_____

MATH SKILLS

5. A bacterial cell infected by a virus divides every 20 min. After 10,000 divisions, the new viruses are released from their host cell. About how many weeks will this process take? Show your work below.

| Section Review *continued*

CRITICAL THINKING

6. Making Inferences Do you think modern transportation has had an effect on the way viruses spread? Explain.

_____

_____

_____

_____

7. Identifying Relationships What characteristics of viruses do you think have made finding drugs to attack them difficult?

_____

_____

_____

_____

8. Expressing Opinions Do you think that vaccinations are important even in areas where a virus is not found?

_____

_____

_____

_____

Skills Worksheet

Chapter Review

USING KEY TERMS

1. In your own words, write a definition for the term *pathogenic bacteria.*

_____

_____

_____

_____

Complete each of the following sentences by choosing the correct term from the word bank.

binary fission	endospore	virus
antibiotic	bioremediation	bacteria

2. Most prokaryotes reproduce by _____.

3. Bacterial infections can be treated with _____.

4. A(n) _____ needs a host to reproduce.

UNDERSTANDING KEY IDEAS
Multiple Choice

_____ **5.** Bacteria are used for all of the following EXCEPT
    a. making certain foods.   **c.** cleaning up oil spills.
    b. making antibiotics.   **d.** preserving fruit.

_____ **6.** In the lytic cycle, the host cell
    a. is destroyed.   **c.** becomes a virus.
    b. destroys the virus.   **d.** undergoes cell division.

_____ **7.** A bacterial cell
    a. is an endospore.   **c.** has a distinct nucleus.
    b. has a loop of DNA.   **d.** is a eukaryote.

_____ **8.** Bacteria
    a. include methane makers.   **c.** all have chlorophyll.
    b. include decomposers.   **d.** are rod-shaped.

_____ **9.** Cyanobacteria
    a. are consumers.   **c.** contain chlorophyll.
    b. are parasites.   **d.** are decomposers.

Chapter Review *continued*

_____10. Archaea
  a. are a special type of eubacteria.
  b. live only in places without oxygen.
  c. are lactic acid-producing bacteria.
  d. can live in hostile environments.

_____11. Viruses
  a. are about the same size as bacteria.
  b. have nuclei.
  c. can reproduce only within a host cell.
  d. do not infect plants.

_____12. Bacteria are important to the planet as
  a. decomposers of dead organic matter.
  b. processors of nitrogen.
  c. makers of medicine.
  d. All of the above

Short Answer

13. How are the functions of nitrogen-fixing bacteria and decomposers similar?

_____

_____

_____

_____

14. Which cycle takes more time, the lytic cycle or the lysogenic cycle?

_____

_____

_____

_____

_____

15. Describe two ways in which viruses do not act like living things.

_____

_____

_____

_____

_____

Chapter Review *continued*

16. What is bioremediation?

_____

_____

_____

_____

17. Describe how doctors can treat a viral infection.

_____

_____

_____

_____

CRITICAL THINKING

18. Concept Mapping Use the following terms to create a concept map:
bacteria, bacilli, cocci, spirilla, consumers, producers, and *cyanobacteria.*

| Chapter Review *continued*

19. Predicting Consequences Describe some of the problems you think bacteria might face if there were no humans.

_____

_____

_____

_____

_____

20. Applying Concepts Many modern soaps contain chemicals that kill bacteria. Describe one good outcome and one bad outcome of the use of antibacterial soaps.

_____

_____

_____

21. Identifying Relationships Some people have digestive problems after they take a course of antibiotics. Why do you think these problems happen?

_____

_____

_____

_____

_____

_____

INTERPRETING GRAPHICS

The diagram below illustrates the stages of binary fission. Match each statement with the correct stage.

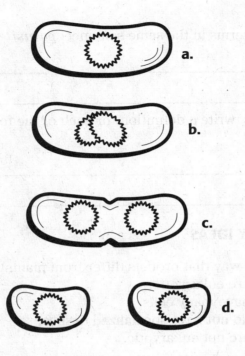

a.

b.

c.

d.

_____22. The DNA loops separate.

_____23. The DNA loop replicates.

_____24. The parent cell starts to expand.

_____25. The DNA attaches to the cell membrane.

Section Review

Protists

USING KEY TERMS

1. Use the following terms in the same sentence: *parasite* and *host*.

_____

_____

2. In your own words, write a definition for each of the following terms: *protist* and *heterotroph*.

_____

_____

UNDERSTANDING KEY IDEAS

_____ 3. What is one way that protists differ from plants and animals?
   a. Protists are eukaryotic.
   b. Protists have many cells.
   c. Protists do not have specialized tissues.
   d. Protists are not eukaryotic.

4. Name a characteristic shared by all protists.

_____

5. Name three ways that protists can differ from each other.

_____

_____

_____

6. Describe four ways that protists get food.

_____

_____

_____

_____

Section Review *continued*

7. Describe three ways that protists reproduce.

_____

_____

_____

MATH SKILLS

8. If seven individuals of the genus *Euglena* reproduce at one time, how many individuals result? Show your work below.

CRITICAL THINKING

9. Identifying Relationships How is conjugation similar to fission?

_____

_____

_____

_____

10. Applying Concepts The spread of malaria depends on both human and mosquito hosts. Use this fact to think of a way to stop the spread of malaria.

_____

_____

_____

_____

Skills Worksheet

Section Review

Kinds of Protists

USING KEY TERMS

1. Use the following terms in the same sentence: *phytoplankton* and *algae*.

_____

_____

UNDERSTANDING KEY IDEAS

_____ 2. Which of the following kinds of protists are producers?
   a. diatoms
   b. amoebas
   c. slime molds
   d. ciliates

_____ 3. How do many amoeba-like protists eat?
   a. They secrete digestive juices onto food.
   b. They produce food from sunlight.
   c. They engulf food with pseudopodia.
   d. They use cilia to sweep food toward them.

4. Give an example of one protist from each of the three groups of protists.

_____

_____

_____

5. Explain why it makes sense to group protists based on shared traits rather than by how they are related to each other.

_____

_____

CRITICAL THINKING

6. **Making Comparisons** How do protist producers, heterotrophs that can move, and heterotrophs that can't move differ?

_____

_____

_____

7. Making Inferences You learned how shelled amoeba-like protists move. How do you think they get food into their shells in order to eat?

_____

_____

_____

INTERPRETING GRAPHICS

Use the photo on the Section Review page in your textbook to answer the questions that follow.

8. How does this protist move?

_____

_____

9. Identify what kind of protist is shown. To do so, first make a list of the kinds of protists that this organism could not be.

_____

_____

_____

_____

Skills Worksheet

Section Review

Fungi

USING KEY TERMS

1. In your own words, write a definition for each of the following terms: *spore* and *mold*.

_____

_____

For each pair of terms, explain how the meanings of the terms differ.

2. *fungus* and *lichen*

_____

_____

3. *hyphae* and *mycelium*

_____

_____

UNDERSTANDING KEY IDEAS

_____ 4. Which of the following statements about fungi is true?
   a. All fungi are eukaryotic.
   b. All fungi are decomposers.
   c. All fungi reproduce by sexual reproduction.
   d. All fungi are producers.

5. What are the four main groups of fungi? Give a characteristic of each group.

_____

_____

_____

_____

6. How are fungi able to withstand periods of cold or drought?

_____

_____

_____

Section Review *continued*

CRITICAL THINKING

7. Analyzing Processes Many fungi are decomposers. Imagine what would happen to the natural world if decomposers no longer existed. Write a description of how a lack of decomposers might affect the processes of nature.

_____

_____

_____

8. Identifying Relationships Explain how two organisms make up a lichen.

_____

_____

_____

INTERPRETING GRAPHICS

Use the drawing below to answer the questions that follow.

9. To which group of fungi does this organism belong? How can you be sure?

_____

_____

_____

10. What part of the organism is shown in this picture? What part is not shown? Explain.

_____

_____

_____

Skills Worksheet

Chapter Review

USING KEY TERMS

1. In your own words, write a definition for each of the following terms: *mycelium*, *lichen*, and *heterotroph*.

_____

_____

_____

2. Use the following terms in the same sentence: *protists*, *algae*, and *phytoplankton*.

_____

_____

_____

3. Use the following terms in the same sentence: *spore* and *mold*.

_____

_____

For each pair of terms, explain how the meanings of the terms differ.

4. *fungus* and *hypha*

_____

_____

5. *parasite* and *host*

_____

_____

UNDERSTANDING KEY IDEAS

Multiple Choice

_____ **6.** Protist producers include
   a. euglenoids and ciliates.
   b. lichens and zooflagellates.
   c. spore-forming protists and smuts.
   d. dinoflagellates and diatoms.

_____ **7.** Protists can be
  a. parasites or decomposers.
  b. made of chains of cells called *hyphae*.
  c. divided into four major groups.
  d. only parasites.

_____ **8.** A euglenoid has
  a. a micronucleus.   **c.** two flagella.
  b. pseudopodia.   **d.** cilia.

_____ **9.** Which statement about fungi is true?
  a. Fungi are producers.
  b. Fungi cannot eat or engulf food.
  c. Fungi are found only in the soil.
  d. Fungi are primarily single celled.

_____ **10.** A lichen is made up of
  a. a fungus and a funguslike protist that live together.
  b. an alga and a fungus that live together.
  c. two kinds of fungi that live together.
  d. an alga and a funguslike protist that live together.

_____ **11.** Heterotrophic protists that can move
  a. are also known as *protozoans*.
  b. include amoebas and paramecia.
  c. may be either free living or parasitic.
  d. All of the above

Short Answer

12. How are fungi helpful to humans?

_____

_____

13. What is the function of cilia in a paramecium?

_____

_____

14. How are fungi different from protists that get food as decomposers?

_____

_____

| **Chapter Review** *continued*

15. How are slime molds and amoebas similar?

_____

_____

16. What is a contractile vacuole?

_____

_____

17. Compare how *Paramecium*, *Plasmodium vivax*, and *Euglena* reproduce.

_____

_____

_____

18. Compare how phytoplankton, amoebas, and *Giardia lamblia* get food.

_____

_____

_____

19. Explain how protists differ from other organisms.

_____

_____

20. Give an example of where you might find each of the following fungi: threadlike fungi, sac fungi, club fungi, and imperfect fungi.

_____

_____

_____

_____

CRITICAL THINKING

21. Concept Mapping Use the following terms to create a concept map: *yeast, basidia, threadlike fungi, mushrooms, fungi, bread mold, ascus,* and *club fungi.*

22. Applying Concepts Why do you think bread turns moldy less quickly when it is kept in a refrigerator than when it is kept at room temperature?

_____

_____

23. Making Inferences Some protozoans, such as radiolarians and foraminiferans, have shells around their bodies. How might these shells be helpful to the protists that live in them?

_____

_____

24. Predicting Consequences Suppose a forest where many threadlike fungi live goes through a very dry summer and fall and then a very cold winter. How could this extreme weather affect the reproductive patterns of these fungi?

_____

_____

INTERPRETING GRAPHICS

Use the pictures of fungi below to answer the questions that follow.

a.
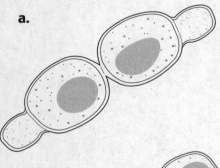

b.
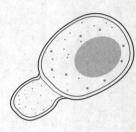

c.
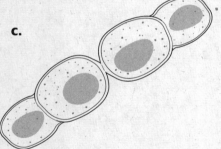

d.
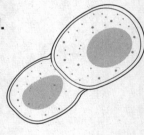

25. What kind of fungus is shown here?

_____

26. What cellular process is shown in these pictures?

_____

27. Which picture was taken first? Which was taken last? Arrange the pictures in order.

_____

_____

28. Which is the original parent cell? How do you know?

_____

_____

Skills Worksheet

Section Review

What Is a Plant?

USING KEY TERMS

For each pair of terms, explain how the meanings of the terms differ.

1. *nonvascular plants* and *vascular plants*

_____

_____

2. *gymnosperms* and *angiosperms*

_____

_____

_____

UNDERSTANDING KEY IDEAS

_____ 3. Which of the following plants is nonvascular?
   a. ferns   **c.** gymnosperms
   b. mosses   **d.** club mosses

4. What are four characteristics that all plants share?

_____

_____

_____

_____

5. What do green algae and plants have in common?

_____

_____

_____

_____

6. Describe the plant life cycle.

_____

_____

_____

_____

| Section Review *continued*

MATH SKILLS

7. A plant produced 200,000 spores and one-third as many eggs. How many eggs did the plant produce? Show your work below.

CRITICAL THINKING

8. Making Inferences One difference between green algae and plants is that green algae do not have a cuticle. Why don't green algae have a cuticle?

_____

_____

_____

_____

_____

_____

_____

9. Applying Concepts Imagine an environment that is very dry and receives a lot of sunlight. Water is found deep below the soil. Which of the four groups of plants could survive in this environment? Explain your answer.

_____

_____

_____

_____

_____

_____

_____

_____

Skills Worksheet

Section Review

Seedless Plants
USING KEY TERMS

1. Use each of the following terms in a separate sentence: *rhizoid* and *rhizome*.

_____

_____

UNDERSTANDING KEY IDEAS

_____ **2.** Seedless plants
  a. help form communities.
  b. reduce soil erosion.
  c. add to soil depth.
  d. All of the above

3. Describe six kinds of seedless plants.

_____

_____

_____

_____

4. What is the relationship between coal and seedless vascular plants?

_____

_____

MATH SKILLS

5. Club mosses once grew as tall as 40 m. Now, they grow no taller than 20 cm. What is the difference in height between ancient and modern club mosses? Show your work below.

Section Review *continued*

CRITICAL THINKING

6. Making Inferences Imagine a very damp area. Mosses cover the rocks and trees in this area. Liverworts and hornworts are also very abundant. What might happen if the area dries out? Explain your answer.

_____

_____

_____

_____

_____

_____

_____

_____

7. Applying Concepts Modern ferns, horsetails, and club mosses are smaller than they were millions of years ago. Why might these plants be smaller?

_____

_____

_____

_____

_____

_____

_____

_____

_____

_____

Section Review

Seed Plants

USING KEY TERMS

1. In your own words, write a definition for each of the following terms:
   pollen and *pollination.*

   _____

   _____

   _____

   _____

UNDERSTANDING KEY IDEAS

_____ 2. One advantage of seed plants is that
   a. seed plants grow in few places.
   b. they can begin photosynthesis as soon as they begin to grow.
   c. they need water for fertilization.
   d. young plants are nourished by food stored in the seed.

_____ 3. The gametophytes of seed plants
   a. live independently of the sporophytes.
   b. are very large.
   c. are protected in the reproductive structures of the sporophyte.
   d. None of the above

4. Describe the structure of seeds.

   _____

   _____

   _____

5. Briefly describe the four groups of gymnosperms. Which group is the largest
   and most economically important?

   _____

   _____

   _____

   _____

| Section Review *continued*

6. Compare angiosperms and gymnosperms.

_____

_____

_____

MATH SKILLS

7. More than 265,000 species of plants have been discovered. Approximately 235,000 of those species are angiosperms. What percentage of plants are NOT angiosperms? Show your work below.

CRITICAL THINKING

8. Making Inferences In what ways are flowers and fruits adaptations that help angiosperms reproduce?

_____

_____

_____

_____

9. Applying Concepts An angiosperm lives in a dense rain forest, close to the ground. It receives little wind. Several herbivores live in this area of the rainforest. What are some ways the plant can ensure its seeds are carried throughout the forest?

_____

_____

_____

_____

_____

Skills Worksheet)

Section Review

Structures of Seed Plants
USING KEY TERMS

1. In your own words, write a definition for each of the following terms: *xylem*, *phloem*, *stamen*, and *pistil*.

_____

_____

_____

_____

2. Use each of the following terms in a separate sentence: *sepal*, *petal*, *pistil*, and *ovary*.

_____

_____

_____

_____

UNDERSTANDING KEY IDEAS

_____ **3.** Which of the following flower structures produces pollen?
    a. pistil   **c.** anther
    b. filament   **d.** stigma

_____ **4.** The _____ of a leaf allows carbon dioxide to enter.
    a. stoma   **c.** palisade layer
    b. epidermis   **d.** spongy layer

5. Compare xylem and phloem.

_____

_____

_____

_____

| Section Review *continued*

6. Describe the internal structure of a leaf.

_____

_____

_____

7. What are the functions of stems?

_____

_____

_____

8. Identify the two types of stems, and briefly describe them.

_____

_____

_____

9. How do people use flowers?

_____

_____

CRITICAL THINKING

10. Making Inferences Describe two kinds of root systems. How does the structure of each system help the roots perform their three functions?

_____

_____

_____

_____

11. **Applying Concepts** Pampas grass flowers are found at the top of tall stems, are light-colored, and are unscented. Explain how pampas flowers are most likely pollinated.

_____

_____

_____

_____

INTERPRETING GRAPHICS

Use the table below to answer the questions that follow.

Age of Trees in a Small Forest	
Number of trees	**Number of growth rings**
5	71
1	73
3	68

12. How many trees are older than 70 years?

_____

13. What is the average age of these trees, in years? Show your work below.

Skills Worksheet

Chapter Review

USING KEY TERMS

Complete each of the following sentences by choosing the correct term from the word bank.

pistil	rhizoid	vascular plant
rhizome	xylem	phloem
pollen	stamen	nonvascular plant

1. A _____ is the male part of a flower.

2. _____ transports water and nutrients through a plant.

3. An underground stem that produces new leaves and roots is called

   a _____.

4. The male gametophytes of flowers are contained in structures

   called _____.

5. A _____ does not have specialized tissues for

   transporting water.

6. _____ transports food through a plant.

UNDERSTANDING KEY IDEAS

Multiple Choice

_____ 7. Which of the following statements about angiosperms is NOT true?
   a. Their seeds are protected by cones.
   b. They produce seeds.
   c. They provide animals with food.
   d. They have flowers.

_____ 8. Roots
   a. supply water and nutrients.
   b. anchor and support a plant.
   c. store surplus food.
   d. All of the above

_____ 9. Which of the following statements about plants and green algae
   is true?
   a. Plants and green algae may have a common ancestor.
   b. Green algae are plants.
   c. Plants and green algae have cuticles.
   d. None of the above

Chapter Review *continued*

_____**10.** In which part of a leaf does most photosynthesis take place?
  a. palisade layer   **c.** xylem
  b. phloem   **d.** epidermis

Short Answer

11. List four characteristics that all plants share.

_____

_____

_____

_____

12. List the four main groups of plants.

_____

_____

_____

_____

13. Name three nonvascular plants and three seedless vascular plants.

_____

_____

_____

_____

14. Why do scientists think green algae and plants have a common ancestor?

_____

_____

15. How are seedless plants, gymnosperms, and angiosperms important to the environment?

_____

_____

16. What are two advantages that seeds have over spores?

_____

_____

_____

Concept Mapping

17. Use the following terms to create a concept map: *flowers, pollen, stamens, ovaries, pistils, stigmas, filaments, anthers, ovules, petals, sepals.*

| Chapter Review *continued*

18. Making Comparisons Imagine that a seed and a spore are beginning to grow in a deep, dark crack in a rock. Which of the two is more likely to grow into an adult plant? Explain your answer.

_____

_____

_____

_____

19. Identifying Relationships Grass flowers do not have strong fragrances or bright colors. How might these characteristics be related to the way by which grass flowers are pollinated?

_____

_____

_____

_____

20. Analyzing Ideas Plants that are pollinated by wind produce more pollen than plants pollinated by animals do. Why might wind-pollinated plants produce more pollen?

_____

_____

_____

_____

21. Applying Concepts A scientist discovered a new plant. The plant has vascular tissue and produces seeds. It has brightly colored and strongly scented flowers. It also has sweet fruits. Based on this information, which of the four main types of plants did the scientist discover? How is the plant most likely pollinated? How does the plant most likely spread its seeds?

_____

_____

_____

_____

Chapter Review *continued*

INTERPRETING GRAPHICS

22. Look at the cross section of a woody stem below. Use the diagram to determine the age of the tree.

_____

_____

Use the diagram of the flower below to answer the questions that follow.

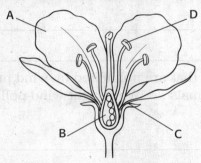

23. Which letter corresponds to the structure in which pollen is produced? What is the name of this structure?

_____

_____

24. Which letter corresponds to the structure that contains the ovules? What is the name of this structure?

_____

_____

25. Which letter corresponds to the structure that protects the flower bud? What is the name of this structure?

_____

_____

Skills Worksheet

Section Review

Photosynthesis

USING KEY TERMS

1. In your own words, write a definition for each of the following terms: *photosynthesis*, *chlorophyll*, and *cellular respiration*.

_____

_____

_____

_____

_____

UNDERSTANDING KEY IDEAS

_____ **2.** During photosynthesis, plants
   a. absorb energy from sunlight.
   b. use carbon dioxide and water.
   c. make food and oxygen.
   d. All of the above

3. How is cellular respiration related to photosynthesis?

_____

_____

_____

_____

4. Describe gas exchange in plants.

_____

_____

_____

_____

Section Review *continued*

MATH SKILLS

5. Plants use 6 carbon dioxide molecules and 6 water molecules to make 1 glucose molecule. How many carbon dioxide and water molecules would be needed to make 12 glucose molecules? Show your work below.

CRITICAL THINKING

6. Predicting Consequences Predict what might happen if plants and other photosynthetic organisms disappeared.

_____

_____

_____

_____

_____

_____

7. Applying Concepts Light filters let through certain colors of light. Predict what would happen if you grew a plant under a green light filter.

_____

_____

_____

_____

_____

_____

_____

Skills Worksheet

Section Review

Reproduction of Flowering Plants

USING KEY TERMS

1. In your own words, write a definition for the term *dormant*.

_____

_____

UNDERSTANDING KEY IDEAS

_____ 2. Pollination happens when
   a. a pollen tube forms.
   b. a sperm cell fuses with an egg.
   c. pollen is transferred from the anther to the stigma.
   d. None of the above

3. Which part of a flower develops into a fruit? into a seed?

_____

_____

4. Why do seeds become dormant?

_____

_____

5. Describe how plants reproduce asexually.

_____

_____

_____

MATH SKILLS

6. A seed sprouts when the temperature is 27°C. If the temperature is now 20°C and it rises 1.5°C per week, in how many weeks will the seed sprout? Show your work below.

| Section Review *continued*

CRITICAL THINKING

7. Making Inferences What do flowers and runners have in common? How do they differ?

_____

_____

_____

_____

_____

_____

8. Identifying Relationships When might asexual reproduction be important for the survival of some flowering plants?

_____

_____

_____

_____

_____

_____

9. Analyzing Ideas Sexual reproduction produces more genetic variety than asexual reproduction. Why is variety important?

_____

_____

_____

_____

_____

_____

Skills Worksheet

Section Review

Plant Responses to the Environment

USING KEY TERMS

1. In your own words, write a definition for the term *tropism*.

_____

_____

UNDERSTANDING KEY IDEAS

_____ **2.** Deciduous trees lose their leaves
   a. to conserve water during the dry season.
   b. around the same time each year.
   c. to survive low winter temperatures.
   d. All of the above

3. How do light and gravity affect plants?

_____

_____

_____

4. Describe how day length can affect the flowering of plants.

_____

_____

_____

MATH SKILLS

5. A certain plant won't bloom until it is dark for 70% of a 24 h period. How long is the day when the plant will bloom? Show your work below.

Section Review *continued*

CRITICAL THINKING

6. Making Inferences Many evergreen trees live in areas with long, cold winters. Why might these evergreen trees keep their leaves all year?

_____

_____

_____

_____

_____

_____

7. Analyzing Ideas Some short-day plants bloom during the winter. If cold weather reduces the chances that a plant will produce seeds, what might you conclude about where these short-day plants are found?

_____

_____

_____

_____

_____

_____

_____

_____

Skills Worksheet

Chapter Review

USING KEY TERMS

Complete each of the following sentences by choosing the correct term from the word bank.

stoma	photosynthesis	dormant
cellular respiration	tropism	chlorophyll
transpiration		

1. The loss of water from leaves is called _____.

2. A plant's response to light or gravity is called a _____.

3. _____ is a green pigment found in plant cells.

4. To get energy from the food made during photosynthesis, plants use

_____.

5. A _____ is an opening in the epidermis and cuticle of a

leaf.

6. An inactive seed is _____.

7. _____ is the process by which plants make their own

food.

UNDERSTANDING KEY IDEAS

Multiple Choice

_____ **8.** During gas exchange in plants,
   a. carbon dioxide exits while oxygen and water enter the leaf.
   b. oxygen and water exit while carbon dioxide enters the leaf.
   c. carbon dioxide and water enter while oxygen exits the leaf.
   d. carbon dioxide and oxygen enter while water exits the leaf.

_____ **9.** Plants often respond to light from one direction by
   a. bending away from the light.
   b. bending toward the light.
   c. wilting.
   d. None of the above

_____ **10.** Which of the following is NOT a way that plants reproduce asexually?
   a. runners
   b. tubers
   c. flowers
   d. plantlets

Chapter Review *continued*

Short Answer

11. Compare short-day plants and long-day plants.

_____
_____
_____

12. How do potted plants respond to gravity if placed on their sides?

_____
_____

13. Describe the pollination and fertilization of flowering plants.

_____
_____
_____
_____
_____

14. What three things do seeds need before they will sprout?

_____
_____

15. Explain how fruits and seeds form from flowers.

_____
_____
_____

16. Compare photosynthesis and cellular respiration.

_____
_____
_____
_____

17. What are two ways in which photosynthesis is important?

_____
_____
_____

CRITICAL THINKING

18. Concept Mapping Use the following terms to create a concept map:
plants, cellular respiration, light energy, photosynthesis, chemical energy, carbon dioxide, and *oxygen*.

Chapter Review *continued*

19. Making Inferences Many plants live in areas that have severe winters. Some of these plants have seeds that will not germinate unless the seeds have first been exposed to a long period of cold. How might this characteristic help new plants survive?

_____

_____

_____

_____

20. Analyzing Ideas Most plant shoots have positive phototropism. Plant roots have positive gravitropism. What might be the benefits of each of these characteristics?

_____

_____

_____

_____

21. Applying Concepts If you wanted to make poinsettias bloom and the leaves turn red in the summer, what would you have to do?

_____

_____

_____

_____

22. Making Inferences Imagine that someone discovered a new flowering plant. The plant has yellow flowers and underground stems. How might this plant reproduce asexually?

_____

_____

_____

_____

Chapter Review *continued*

INTERPRETING GRAPHICS

The graph below shows seed germination rates for different seed companies. Use the graph below to answer the questions that follow.

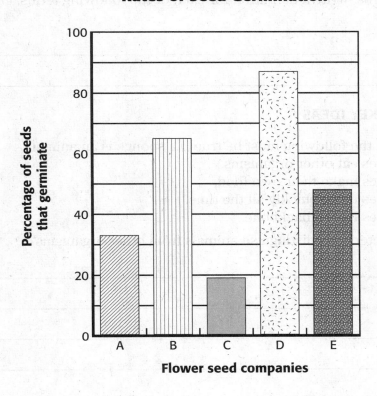

Rates of Seed Germination

Flower seed companies

23. Which seed company had the highest rate of seed germination? the lowest rate of seed germination?

_____

24. Which seed companies had seed germination rates higher than 50%?

_____

25. If Elaine wanted to buy seeds that had a germination rate higher than 60%, which seed companies would she buy seeds from? Why might Elaine want to buy seeds with a higher germination rate?

_____

_____

_____

_____

Section Review

What Is an Animal?

USING KEY TERMS

1. In your own words, write a definition for each of the following terms: *embryo* and *consumer*.

_____

_____

UNDERSTANDING KEY IDEAS

_____ **2.** Which of the following must be true if a sponge is an animal?
  a. Sponges eat other organisms.
  b. Sponges make their own food.
  c. Sponges move quickly all the time.
  d. Sponges have a backbone.

3. What five characteristics distinguish animals from other organisms?

_____

_____

_____

_____

_____

4. How are vertebrates different from invertebrates?

_____

_____

MATH SKILLS

5. If a fish can swim short distances at 48 km/h, how long would the fish take to reach a smaller fish that is 3 m away? Show your work below.

| Section Review *continued*

CRITICAL THINKING

6. Applying Concepts Choose an animal that interests you. Explain how you know that this organism is an animal.

_____

_____

_____

_____

_____

7. Identifying Relationships Suppose that a certain fish tank contains the following: water, chemicals, fish, snails, algae, plants, and gravel. Which of these items are alive? Which are animals? Why aren't some of the living organisms classified as animals?

_____

_____

_____

_____

_____

_____

_____

_____

Skills Worksheet

Section Review

Animal Behavior
USING KEY TERMS

1. Use each of the following terms in a separate sentence: *territory*, *innate behavior*, and *circadian rhythm*.

_____

_____

_____

2. In your own words, write a definition for each of the following terms: *hibernation* and *estivation.*

_____

_____

UNDERSTANDING KEY IDEAS

_____ 3. An animal that lives in a hot, dry environment might spend the summer
   a. hibernating.
   b. estivating.
   c. migrating to a warmer climate.
   d. None of the above

_____ 4. Biological clocks control
   a. seasonal cycles.
   b. circadian rhythms.
   c. internal cycles.
   d. All of the above

5. How do innate behaviors and learned behaviors differ?

_____

_____

6. Do bears hibernate? Explain your answer.

_____

_____

Section Review *continued*

7. Name five behaviors that help animals survive.

_____

_____

_____

_____

_____

MATH SKILLS

8. Suppose that an animal's circadian rhythms tell it to eat a meal every 4 h. How many meals will the animal eat each day? Show your work below.

CRITICAL THINKING

9. Applying Concepts People who travel to different time zones often suffer from *jet lag*. Jet lag makes people have trouble waking up and going to sleep at appropriate times. Why do you think people experience jet lag? Explain.

_____

_____

_____

10. Making Inferences Many children are born with the tendency to make babbling sounds. But few adults make these sounds. How could you explain this change in an innate behavior?

_____

_____

_____

_____

Skills Worksheet

Section Review

Social Relationships
USING KEY TERMS

1. Use each of the following terms in a separate sentence: *social behavior* and *communication*.

_____

_____

2. In your own words, write a definition for the following term: *pheromone*.

_____

_____

_____

UNDERSTANDING KEY IDEAS

_____ **3.** Which of the following is NOT an example of social behavior?
  a. a wolf howling at distant wolves to protect its territory
  b. a rabbit hiding from a predator
  c. a ground squirrel calling to signal danger to other squirrels
  d. a group of lions working together to hunt prey

4. Describe four ways that animals communicate with each other. Give an example of each type of communication.

_____

_____

_____

_____

5. Compare the costs and benefits of living in a group of animals.

_____

_____

_____

_____

_____

MATH SKILLS

6. How fast could a bee that flies 6 km/h reach a flower that is 1.2 km from the hive? Show your work below.

CRITICAL THINKING

7. Applying Concepts Why do you think humans live together?

_____

_____

_____

_____

_____

_____

8. Identifying Relationships Language is not the only way that humans communicate. Describe how we use sound, touch, chemicals, and sight to communicate.

_____

_____

_____

_____

_____

_____

_____

Skills Worksheet

Chapter Review

USING KEY TERMS

1. In your own words, write a definition for each of the following terms: *embryo*, *consumer*, and *pheromone*.

_____

_____

_____

2. Use the following terms in the same sentence: *estivation*, *hibernation*, and *circadian rhythm*.

_____

_____

_____

For each pair of terms, explain how the meanings of the terms differ.

3. *social behavior* and *communication*

_____

_____

4. *learned behavior* and *innate behavior*

_____

_____

UNDERSTANDING KEY IDEAS

Multiple Choice

_____ 5. Which of the following is a characteristic of all animals?
   a. asexual reproduction
   b. producing their own food
   c. having many specialized parts
   d. being unable to move

_____ 6. An innate behavior
   a. cannot change.
   b. must be learned from parents.
   c. is always present from birth.
   d. does not depend on learning or experience.

Chapter Review *continued*

_____ **7.** Migration
   a. occurs only in birds.
   b. helps animals escape cold and food shortages in winter.
   c. always refers to moving southward for the winter.
   d. is a way to defend against predators.

_____ **8.** A biological clock controls
   a. circadian rhythms.   **c.** learned behavior.
   b. defensive behavior.   **d.** being a consumer.

_____ **9.** For animals, living as part of a group
   a. is always safer than living alone.
   b. can attract attention from predators.
   c. keeps them from killing large prey.
   d. decreases competition for mates and food.

Short Answer

10. What is a territory? Give an example of a territory from your environment.

_____

_____

11. What landmarks help you find your way home from school?

_____

_____

12. What are five behaviors that animals may use to survive?

_____

_____

_____

_____

13. What do migration and hibernation have in common?

_____

_____

14. Describe the differences between vertebrates and invertebrates.

_____

_____

Chapter Review *continued*

15. Describe four ways that an animal could communicate a message to other animals about where to find food.

_____

_____

_____

_____

CRITICAL THINKING

16. Concept Mapping Use the following terms to create a concept map: *animals, survival behavior, finding food, migration, defensive action, seasonal behavior, marking a territory, estivation, parenting, hibernation,* and *courtship.*

Chapter Review *continued*

17. Analyzing Processes If you see a skunk raise its tail toward you while you are hiking and you turn around to take a different path, has the skunk communicated with you? Explain your answer.

_____

_____

_____

_____

18. Making Inferences Ants depend on pheromones and touch for communication, but birds depend more on sight and sound. Why might these two types of animals have different forms of communication?

_____

_____

_____

_____

19. Making Comparisons Dogs use visual communication in many situations. They may arch their back and raise their fur to look threatening. When they want to play, they may bow down on their front legs. How are these two visual signals different from each other? How do the different visual signals relate to the different information they are meant to communicate?

_____

_____

_____

_____

20. Analyzing Ideas People have internal biological clocks. However, people are used to keeping track of time by using clocks and calendars. Why do you think people use these tools if they have internal clocks?

_____

_____

_____

_____

21. Applying Concepts Imagine that you are taking care of a friend's cat for a few days but that the friend forgot to tell you where to find the cat food. When you arrive at the friend's house, the cat meows and runs to the door that leads to the garage. Where would you look for the cat food? What kind of communication led you to this conclusion?

_____

_____

_____

INTERPRETING GRAPHICS

The diagram below shows some internal organs of a fish. Use the diagram below to answer the questions that follow.

22. What characteristics suggest that this organism is an animal?

_____

_____

23. Which labels point to the animal's organs? Name any organs that you can recognize.

_____

_____

24. Do any labels point to the animal's tissues? Explain.

_____

_____

25. Is this animal a vertebrate or an invertebrate? Explain.

_____

_____

Section Review

Simple Invertebrates
USING KEY TERMS

Complete each of the following sentences by choosing the correct term from the word bank.

invertebrate	gut
ganglion	coelom

1. A(n) _____ is a mass of nerve cells that controls an

animal's actions.

2. A(n) _____ does not have a backbone.

3. The _____ is a special space in an animal's body that

stores the_____ and other organs.

UNDERSTANDING KEY IDEAS

_____ **4.** Which of the following is a trait shared by all invertebrates?
   a. having no backbone
   b. having radial symmetry
   c. having a brain
   d. having a gut

_____ **5.** What do sponges use to digest food?
   a. an osculum
   b. pores
   c. collar cells
   d. a gut

6. Describe cnidarian body forms and stinging cells.

_____

_____

_____

7. How is a roundworm similar to a piece of spaghetti?

_____

_____

_____

Name _____ Class _____ Date _____

INTERPRETING GRAPHICS

All invertebrate nervous systems are made up of some or all of the same basic parts. The drawing below shows the nervous system of a segmented worm. Use this drawing to answer the questions that follow.

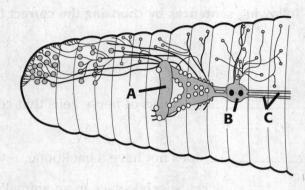

8. The letters in the drawing point to nerve cords, a ganglion, and a brain. Which letter points to the brain? How can you tell?

_____

_____

_____

_____

9. How is the brain connected to the ganglion?

_____

_____

_____

_____

CRITICAL THINKING

10. **Making Inferences** Explain why it would be important for a parasite that its host survive.

_____

_____

_____

_____

Name _____ Class _____ Date _____

Skills Worksheet)

Section Review

Mollusks and Annelid Worms
USING KEY TERMS

1. Use the following terms in the same sentence: *open circulatory system* and *closed circulatory system.*

_____

_____

_____

2. In your own words, write a definition for the term *segment.*

_____

_____

UNDERSTANDING KEY IDEAS

_____ **3.** Some mollusks use a radula to
   a. scrape algae off rocks.
   b. filter food from water.
   c. grab food from the water.
   d. place food in their jaws.

4. What trait do all mollusk nervous systems share? What is unique about squids' and octopuses' nervous systems?

_____

_____

_____

5. What are the four main body parts of most mollusks?

_____

_____

_____

Section Review *continued*

6. Describe three different kinds of annelid worms.

_____

_____

_____

_____

MATH SKILLS

7. If a squid swims at 30 km/h, how far can it swim in 1 min? Show your work below.

CRITICAL THINKING

8. Predicting Consequences Clams use gills to filter food from water. How could water pollution affect clams?

_____

_____

_____

9. Analyzing Ideas Cephalopods do not have shells. What other traits do they have to help make up for this lack of protection?

_____

_____

_____

_____

_____

Section Review

Arthropods
USING KEY TERMS

1. Use the following terms in the same sentence: *compound eye* and *antenna*.

_____

_____

2. In your own words, write a definition for each of the following terms: *exoskeleton* and *metamorphosis*.

_____

_____

_____

_____

UNDERSTANDING KEY IDEAS

_____ **3.** Which of the following is NOT a trait shared by all arthropods?
   a. exoskeleton
   b. body segments
   c. antennae
   d. jointed limbs

_____ **4.** Which of these arthropods is an arachnid?
   a. butterfly
   b. tick
   c. centipede
   d. lobster

5. What is the difference between complete metamorphosis and incomplete metamorphosis?

_____

_____

_____

_____

Section Review *continued*

6. Name the four kinds of arthropods. How do their bodies differ?

_____

_____

_____

7. Which arthropods have chelicerae? Which have mandibles?

_____

_____

MATH SKILLS

8. How many segments does a millipede with 752 legs have? How many segments does a centipede with 354 legs have? Show your work below.

CRITICAL THINKING

9. Applying Concepts Suppose that you find an arthropod in a swimming pool. The organism has compound eyes, antennae, and wings. Is it a crustacean? Why or why not?

_____

_____

_____

10. Forming Hypotheses Suppose you have found several cocoons on a plant outside your school. Develop a hypothesis about what animal is inside the cocoon. How could you find out if your hypothesis is correct?

_____

_____

_____

_____

Skills Worksheet

Section Review

Echinoderms
USING KEY TERMS

1. Use each of the following terms in a separate sentence: *endoskeleton* and *water vascular system.*

_____

_____

_____

_____

UNDERSTANDING KEY IDEAS

_____ **2.** Which of the following is NOT a trait found in echinoderms?
   a. an endoskeleton
   b. spiny skin
   c. a water vascular system
   d. a nerve ring

3. What is the path taken by water as it flows through the parts of the water vascular system?

_____

_____

_____

_____

4. How are sea cucumbers different from other echinoderms?

_____

_____

_____

_____

5. How does an echinoderm's body symmetry change with age?

_____

_____

| Section Review *continued*

6. Name five different classes of echinoderms. List at least one trait for each group.

_____

_____

_____

_____

_____

MATH SKILLS

7. A sea lily lost 12 of its 178 arms in a hurricane. What percentage of its arms were NOT damaged? Show your work below.

CRITICAL THINKING

8. Making Comparisons How are echinoderms different from and similar to other invertebrates?

_____

_____

_____

9. Making Inferences Suppose you found a sea star with four long arms and one short arm. What might explain the difference?

_____

_____

_____

_____

Skills Worksheet

Chapter Review

USING KEY TERMS

1. In your own words, write a definition for each of the following terms:
ganglion, water vascular system, and *coelom.*

_____

_____

_____

_____

2. Use the following terms in the same sentence: *open circulatory system*
and *invertebrate.*

_____

_____

_____

Complete each of the following sentences by choosing the correct term from the word bank.

| antennae | exoskeleton | coelom | gut |
| compound eyes | metamorphosis | endoskeleton | segments |

3. Almost all invertebrates digest food in a(n) _____.

4. Repeating _____ make up the bodies of annelid worms
and arthropods.

5. A crab's _____ keeps it from losing water.

6. Arthropods use _____ to see images.

7. Echinoderms have spines on their _____.

8. Arthropods use _____ to touch, taste, and smell.

9. Insects change form during _____.

UNDERSTANDING KEY IDEAS
Multiple Choice

_____**10.** No invertebrates have
    a. a brain.   **c.** ganglia.
    b. a gut.   **d.** a backbone.

_____11. Which animals have a nerve ring?
  a. sponges
  b. echinoderms
  c. crustaceans
  d. flatworms

_____12. Which of the following is NOT a flatworm?
  a. a tapeworm
  b. an earthworm
  c. a planarian
  d. a fluke

_____13. Which body part is NOT present in all mollusks?
  a. foot
  b. visceral mass
  c. mantle
  d. shell

Short Answer

14. Describe how a sponge eats.

_____

_____

_____

_____

15. What are the four main characteristics of arthropods?

_____

_____

16. Describe the body of a roundworm.

_____

_____

17. What are three ways that different mollusks eat?

_____

_____

_____

18. Which insects go through complete metamorphosis? go through incomplete metamorphosis?

_____

_____

19. How is an adult echinoderm different from an echinoderm larva?

_____

Chapter Review *continued*

20. How are cephalopod nervous systems unique among mollusks?

_____

_____

CRITICAL THINKING

21. Concept Mapping Use the following terms to create a concept map:
*segments, invertebrates, endoskeleton, antennae, exoskeleton,
water vascular system, metamorphosis,* and *compound eyes.*

22. Applying Concepts You have discovered a new animal that has radial symmetry and tentacles with stinging cells. Can this animal be classified as a cnidarian? Explain.

_____

_____

_____

_____

23. Making Inferences Unlike other mollusks, cephalopods can move quickly. Based on what you know about the structure and function of mollusks, why do you think that cephalopods have this ability?

_____

_____

_____

_____

24. Making Comparisons Why don't roundworms, flatworms, and annelid worms belong to the same group of invertebrates?

_____

_____

_____

_____

25. Analyzing Processes Butterflies mate as adults and spend time eating and growing in their other stages. They have no wings during the larval or pupal stage of metamorphosis. Can you think of a reason that they would need wings in their adult form more than in the other stages of development? Explain your answer.

_____

_____

_____

_____

Chapter Review *continued*

26. Predicting Consequences How do earthworms affect gardens? What do you think would happen to a garden if the gardener removed all the earthworms from it?

_____

_____

_____

_____

INTERPRETING GRAPHICS

The picture below shows an arthropod. Use the picture to answer the questions that follow.

27. Name the body segments labeled a, b, and c.

_____

_____

28. How many legs does this arthropod have?

_____

_____

29. To which segment are the arthropod's legs attached?

_____

_____

30. What kind of arthropod is this?

_____

_____

Skills Worksheet

Section Review

Fishes: The First Vertebrates

USING KEY TERMS

1. Use each of the following terms in a separate sentence: *vertebrate*, *lateral line system*, *gill*, and *swim bladder*.

_____

_____

_____

_____

2. In your own words, write a definition for each of the following terms: *endotherm* and *ectotherm*.

_____

_____

UNDERSTANDING KEY IDEAS

_____ 3. At some point in its life, every chordate has each of the following EXCEPT
    a. a tail.
    b. a notochord.
    c. a hollow nerve cord.
    d. a backbone.

4. Which vertebrates are ectotherms?

_____

_____

5. What are four characteristics shared by most fishes?

_____

_____

_____

_____

| Section Review *continued*

6. What are the three classes of living fish? Give an example of each.

_____

_____

_____

7. Most bony fishes reproduce by external fertilization. What does this mean?

_____

_____

CRITICAL THINKING

8. Analyzing Relationships Describe the ways that cartilaginous fishes and bony fishes maintain buoyancy. Why do you think that jawless fishes do not use one of these methods?

_____

_____

_____

9. Applying Concepts How could moving a fishbowl from a cold window sill to a warmer part of the house affect a pet fish?

_____

_____

_____

| Section Review *continued*

INTERPRETING GRAPHICS

Use the bar graph below to answer the questions that follow.

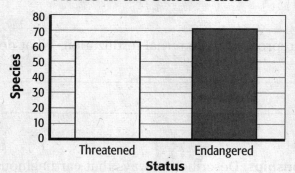

Fishes in the United States

Source: U.S. Fish and Wildlife Services

10. How many fish species in the United States are threatened? How many are endangered?

_____

_____

11. What is the total number of threatened and endangered fish species in the United States?

_____

_____

Skills Worksheet

Section Review

Amphibians

USING KEY TERMS

1. Use each of the following terms in a separate sentence: *lung, tadpole,* and *metamorphosis.*

_____

_____

_____

UNDERSTANDING KEY IDEAS

_____ 2. The first vertebrates to live on land were
   a. fish.
   b. dinosaurs.
   c. amphibians.
   d. reptiles.

_____ 3. Many adult amphibians breathe by using
   a. only their gills.
   b. only their lungs.
   c. only their skin.
   d. their lungs and skin.

4. Describe metamorphosis in amphibians.

_____

_____

5. Why do adult amphibians have to live near water or in a very wet habitat?

_____

_____

6. Why are amphibians sometimes called *ecological indicators*?

_____

_____

7. Name the three types of amphibians. How are they similar? How are they different?

_____

_____

_____

| Section Review *continued* |

8. How are frogs and toads similar? How are they different?

_____

_____

MATH SKILLS

9. A certain toad species spends 2 months of its life as a tadpole and 3 years of its life as an adult. What percentage of its life is spent in the water? What percentage is spent on land? Show your work below.

CRITICAL THINKING

10. Analyzing Relationships Describe the relationship between lungfishes and amphibians. How are these animals alike? How are they different?

_____

_____

_____

_____

11. Evaluating Conclusions Scientists think that climate change may have caused the golden toad to become extinct. What other causes are possible, and how could scientists test these ideas?

_____

_____

_____

_____

Skills Worksheet

Section Review

Reptiles

USING KEY TERMS

1. In your own words, write a definition for the term *amniotic egg.*

_____

_____

UNDERSTANDING KEY IDEAS

_____ 2. Reptiles are well adapted to living on land because they
   a. have thick, scaly skin.
   b. have lungs.
   c. lay amniotic eggs.
   d. All of the above

_____ 3. A reptile can lay its egg on land because
   a. the egg's shell prevents fertilization.
   b. the egg's shell keeps moisture inside the egg.
   c. the egg's shell keeps carbon dioxide inside the egg.
   d. the egg's shell allows water to leave the egg.

4. Name three ways that an amniotic egg protects reptile embryos.

_____

_____

_____

5. Explain how most reptiles reproduce.

_____

_____

6. Name the four groups of modern reptiles, and give an example of each kind.

_____

_____

_____

_____

| Section Review continued

7. What special adaptations do snakes have for eating?

_____

_____

MATH SKILLS

8. Suppose that a sea turtle lays 104 eggs. If 50% of the hatchlings reach the ocean alive and 25% of those survivors reach adulthood, how many adults result from the eggs? Show your work below.

CRITICAL THINKING

9. Applying Concepts Mammals give birth to live young. The embryo develops inside the female's body. Which parts of a reptilian amniotic egg could a mammal do without? Explain.

_____

_____

_____

_____

10. Analyzing Ideas Rattlesnakes can't see well, but they can detect temperature changes of three-thousandths of a degree Celsius. How could this ability be useful to the snakes?

_____

_____

_____

_____

_____

Chapter Review

USING KEY TERMS

1. In your own words, write a definition for each of the following terms: *metamorphosis*, *amniotic egg*, and *vertebrate*.

_____

_____

_____

2. Use the following terms in the same sentence: *lung*, *gills*, and *tadpole*.

_____

_____

For each pair of terms, explain how the meanings of the terms differ.

3. *endotherm* and *ectotherm*

_____

_____

4. *swim bladder* and *lateral line*

_____

_____

UNDERSTANDING KEY IDEAS

Multiple Choice

_____ **5.** Which of the following structures is not present in some chordates?
  a. a tail
  b. a backbone
  c. a notochord
  d. a hollow nerve cord

_____ **6.** Which fishes do not have jaws?
  a. sharks, skates, and rays
  b. hagfish and lampreys
  c. bony fishes
  d. None of the above

_____ **7.** Both amphibians and reptiles
  a. have lungs.
  b. have gills.
  c. breathe only through their skin.
  d. have amniotic eggs.

| Chapter Review *continued*

_____ **8.** Metamorphosis occurs in
   a. fishes and amphibians.
   b. amphibians.
   c. fishes, amphibians, and reptiles.
   d. amphibians and reptiles.

_____ **9.** Both bony fishes and cartilaginous fishes have
   a. fins.
   b. an oily liver.
   c. a swim bladder.
   d. skeletons made of bone.

Short Answer

10. How do amphibians breathe?

_____

_____

11. What characteristics allow fishes to live in the water?

_____

_____

12. What characteristics allow reptiles to live on land?

_____

_____

13. How does a reptile embryo in an amniotic egg get oxygen?

_____

_____

14. Describe the stages of metamorphosis in a frog.

_____

_____

_____

_____

15. What two things are present in all vertebrates but not in some chordates?

_____

_____

| Chapter Review *continued*

16. Describe the three kinds of amphibians.

_____

_____

_____

17. Explain why amphibians can be effective ecological indicators.

_____

_____

CRITICAL THINKING

18. Concept Mapping Use the following terms to create a concept map: *dinosaur, turtle, reptiles, amphibians, fishes, shark, salamander,* and *vertebrates*.

19. Applying Concepts If the air temperature outside is 43°C and the ideal body temperature of a lizard is 38°C, would you most likely find that lizard in the sun or in the shade? Explain your answer.

_____

_____

_____

20. Identifying Relationships Describe three characteristics of amphibian skin. How do amphibians use their skin? How does the structure of amphibian skin relate to its function?

_____

_____

_____

21. Making Inferences Suppose that you have found an animal that has a backbone and gills, but the animal does not seem to have a notochord. Is the animal a chordate? How can you be sure of your answer?

_____

_____

_____

22. Analyzing Processes If you found a shark that lacks the muscles needed to pump water over its gills, what would that information tell you about how the shark lives?

_____

_____

_____

23. Forming Hypotheses If you found a reptile that you did not recognize, what questions would you need to ask to determine which of the four reptile groups the reptile belongs to? Explain how you could form a hypothesis about the reptile's group based on the answers to these questions.

_____

_____

_____

Name _____ Class _____ Date _____

INTERPRETING GRAPHICS

The graph below shows body temperatures of two organisms and the ground temperature of their environment. Use the graph to answer the questions that follow.

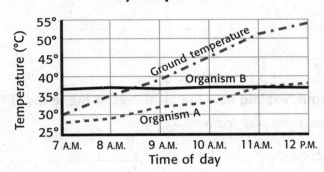

Body Temperatures

24. How do the body temperatures of organism A and organism B change as the ground temperature changes?

_____

_____

_____

_____

25. Which of these organisms is most likely an ectotherm? Explain your answer.

_____

_____

26. Which of these organisms is most likely an endotherm? Explain your answer.

_____

_____

Skills Worksheet

Section Review

Characteristics of Birds

USING KEY TERMS

1. Use each of the following terms in a separate sentence: *lift* and *brooding*.

_____

_____

For each pair of terms, explain how the meanings of the terms differ.

2. *down feather* and *contour feather*

_____

_____

3. *preening* and *molting*

_____

_____

UNDERSTANDING KEY IDEAS

_____ **4.** Which of the following is NOT a flight adaptation in birds?
- **a.** hollow bones
- **b.** air sacs
- **c.** down feathers
- **d.** rapidly beating heart

5. What do birds eat? Describe the path taken by a bird's food as it moves through the bird's digestive system.

_____

_____

6. How does the air around a bird's wing cause lift?

_____

_____

7. Explain the difference between precocial chicks and altricial chicks.

_____

_____

Section Review *continued*

8. Name two ways that birds use their contour feathers. Name one way that birds use their down feathers.

_____

_____

_____

MATH SKILLS

9. Suppose that a bird that weighs 325 g loses 40% of its body weight during migration. What is the bird's weight when it reaches its destination? Show your work below.

CRITICAL THINKING

10. Analyzing Ideas Why can't people fly without the help of technology? Name at least four human body characteristics that are poorly adapted for flight.

_____

_____

_____

_____

11. Applying Concepts Some people use the phrase "eats like a bird" to describe someone who does not eat very much. Does using the phrase in this way show an accurate understanding of a bird's eating habits? Why or why not?

_____

_____

_____

_____

Skills Worksheet)

Section Review

Kinds of Birds
UNDERSTANDING KEY IDEAS

_____ 1. Which of the following groups of birds includes birds that do NOT
have a large keel?
a. flightless birds
b. water birds
c. perching birds
d. birds of prey

_____ 2. Why do some water birds have long legs?
a. for swimming
b. for wading
c. for running
d. for flying

3. Most birds of prey have very good eyesight. Why do you think good vision is
important for these birds?

_____

_____

4. To which group of birds do songbirds belong? Name three examples of
songbirds.

_____

_____

_____

MATH SKILLS

5. How quickly could an ostrich, running at a speed of 60 km/h, run a 400 m
track event? Show your work below.

Section Review *continued*

CRITICAL THINKING

6. Predicting Consequences Would it be helpful for a duck to have the feet of a perching bird? Explain why or why not.

_____

_____

_____

_____

7. Making Inferences How could being able to run 64 km/h be helpful for an ostrich?

_____

_____

_____

_____

Skills Worksheet

Section Review

Characteristics of Mammals

USING KEY TERMS

1. Use each of the following terms in a separate sentence: *mammary gland* and *diaphragm*.

_____

_____

UNDERSTANDING KEY IDEAS

_____ 2. Large brains help mammals survive by allowing them
   a. to think and learn quickly.
   b. to maintain their body temperature.
   c. to have hair all over their body.
   d. to depend on all of the senses equally.

3. What does a diaphragm do?

_____

_____

4. Name three characteristics that are unique to mammals.

_____

_____

_____

5. Describe three characteristics that help mammals stay warm.

_____

_____

_____

6. How are mammal teeth different from reptile and fish teeth?

_____

_____

7. How do mammals reproduce?

_____

_____

Section Review *continued*

MATH SKILLS

8. What is the mass of a 90,000 kg whale in grams? in milligrams? Show your work below.

CRITICAL THINKING

9. Making Inferences Early endothermic mammals could be active at night. If this protected them from certain dinosaurs, were the dinosaurs endothermic? Explain.

_____

_____

_____

10. Applying Concepts How could the teeth of a skull give you clues about a mammal's diet?

_____

_____

_____

Skills Worksheet

Section Review

Placental Mammals

USING KEY TERMS

1. Use the following terms in the same sentence: *placental mammal* and *gestation period*.

_____

_____

UNDERSTANDING KEY IDEAS

_____ 2. Which mammals live entirely in the water?
   a. manatees, dugongs, cetaceans, and pinnipeds
   b. only manatees and dugongs
   c. only cetaceans
   d. manatees, dugongs, and cetaceans

_____ 3. A placental mammal's embryo
   a. develops in the uterus.
   b. develops in the placenta.
   c. develops in a pouch.
   d. develops in a leathery egg.

4. Could you tell a horse from a deer just by looking at their feet? Explain.

_____

_____

5. Give one example of each type of placental mammal described in the section.

_____

_____

_____

_____

_____

Section Review *continued*

CRITICAL THINKING

6. Making Inferences What is a gestation period? Why do you think elephants have a longer gestation period than mice do?

_____

_____

_____

7. Identifying Relationships Manatees may look a little like pinnipeds, but they are more closely related to elephants. In what ways is a manatee more like an elephant than a pinniped?

_____

_____

_____

INTERPRETING GRAPHICS

Use the picture of the animal shown in your textbook for this Section Review to answer the following questions.

8. To which placental mammal group does this animal belong? How can you tell?

_____

_____

9. Why can't this animal be a rodent?

_____

_____

10. Why can't this animal be a primate?

_____

_____

Skills Worksheet

Section Review

Monotremes and Marsupials
USING KEY TERMS

1. Use each of the following terms in a separate sentence: *monotreme* and *marsupial*.

_____

_____

UNDERSTANDING KEY IDEAS

_____ 2. Which of the following characteristics is shared by monotremes and marsupials?
   a. The young hatch from eggs.
   b. Some species of both live in South America.
   c. Females have no nipples.
   d. Females produce milk.

3. What are the two kinds of monotremes?

_____

_____

4. Name three kinds of marsupials.

_____

_____

_____

5. What has caused many marsupials in Australia to become endangered or extinct?

_____

_____

_____

6. How are monotremes different from all other mammals? How are they similar?

_____

_____

_____

MATH SKILLS

7. What percentage of the approximately 5,000 known species of mammals are monotremes? Show your work below.

CRITICAL THINKING

8. Making Comparisons How are monotremes similar to birds? How are they different?

_____

_____

_____

_____

9. Making Inferences Why do you think opossums play dead when they are in danger?

_____

_____

_____

_____

Skills Worksheet

Chapter Review

USING KEY TERMS

1. Use the following terms in the same sentence: *mammary gland*, *placental mammal*, *marsupial*, and *monotreme*.

_____

_____

_____

Complete each of the following sentences by choosing the correct term from the word bank.

brooding	gestation period	contour feathers
lift	diaphragm	molting
down feathers	preening	

2. The _____ is a muscle that helps animals breathe.

3. The embryos of placental mammals develop during

a _____.

4. Birds grow new feathers as a part of the _____ process.

5. _____ help keep birds warm by trapping air near the body.

6. Birds use the _____ process to keep their eggs warm.

7. _____ form a streamlined surface that helps birds fly.

UNDERSTANDING KEY IDEAS
Multiple Choice

_____ **8.** Both birds and reptiles
　　a. lay eggs.
　　b. brood their young.
　　c. have air sacs.
　　d. have feathers.

_____ **9.** Only mammals
　　a. use internal fertilization.
　　b. nurse their young.
　　c. lay eggs.
　　d. have teeth.

_____ **10.** Which of the following is NOT a primate?
　　a. a lemur
　　b. a human
　　c. a pika
　　d. a chimpanzee

Chapter Review *continued*

_____ **11.** Monotremes do NOT
    a. have mammary glands.
    b. care for their young.
    c. give birth to live young.
    d. have hair.

_____ **12.** What is lift?
    a. air that travels over the top of a wing
    b. a force provided by a bird's air sacs
    c. the upward force on a wing that keeps a bird in the air
    d. a force created by pressure from the diaphragm

Short Answer

13. How are contour feathers and down feathers helpful to birds?

_____

_____

_____

14. How do flightless birds, water birds, perching birds, and birds of prey differ from each other?

_____

_____

_____

15. Which trait allowed early mammals to look for food at night?

_____

_____

16. Describe two ways that animals introduced to Australia threaten its native marsupials.

_____

_____

_____

17. Which kind of marsupial lives in North America?

_____

18. Which group of placental mammals includes the pinnipeds?

_____

19. How is a bird's digestive system related to its ability to fly?

_____

_____

20. How can mammalian milks differ?

_____

_____

_____

CRITICAL THINKING

21. Concept Mapping Use the following terms to create a concept map: *monotremes, endotherms, birds, mammals, mammary glands, placental mammals, marsupials, feathers,* and *hair.*

Chapter Review *continued*

22. Making Comparisons The embryos of birds and monotremes get energy from the yolk of the egg. How do developing embryos of marsupials and placental mammals get the nutrition they need?

_____

_____

_____

_____

23. Making Inferences Most bats and cetaceans use echolocation. Why don't these mammals rely solely on sight to hunt and sense their surroundings?

_____

_____

_____

24. Applying Concepts Suppose you are making a museum display of bird skeletons, but the skeletons have lost their labels. How can you separate the skeletons of flightless birds from those of birds that fly? Will you be able to tell which birds flew rapidly and which birds could soar? Explain your answer.

_____

_____

_____

_____

25. Making Inferences Suppose that you saw a bird flying above you. The bird has long, skinny legs and a long, sharp beak. To which group of birds do you think this bird probably belongs? Explain your answer.

_____

_____

_____

_____

_____

Chapter Review *continued*

INTERPRETING GRAPHICS

The illustrations below show three different kinds of bird feet. Use these illustrations to answer the questions that follow.

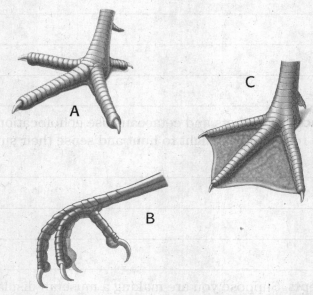

26. Which foot most likely belongs to a water bird? Explain your choice.

_____

_____

_____

27. Which foot most likely belongs to a perching bird? Explain your choice.

_____

_____

_____

28. To what kind of bird do you think the remaining foot could belong? Explain your answer.

_____

_____

_____

Skills Worksheet

Section Review

Everything Is Connected
USING KEY TERMS

1. In your own words, write a definition for the term *ecology*.

_____

_____

2. Use the following terms in the same sentence: *biotic* and *abiotic*.

_____

_____

UNDERSTANDING KEY IDEAS

_____ **3.** Which one of the following is the highest level of environmental organization?
    a. ecosystem   **c.** population
    b. community   **d.** organism

4. What makes up a community?

_____

_____

5. Give two examples of how abiotic factors can affect an ecosystem.

_____

_____

MATH SKILLS

6. From sea level, the biosphere goes up about 9 km and down about 19 km. What is the thickness of the biosphere in meters? Show your work below.

CRITICAL THINKING

7. Analyzing Relationships What would happen to the other organisms in the salt-marsh ecosystem if the cordgrass suddenly died?

_____

_____

_____

_____

8. Identifying Relationships Explain in your own words what people mean when they say that everything is connected.

_____

_____

_____

_____

9. Analyzing Ideas Do ecosystems have borders? Explain your answer.

_____

_____

_____

_____

Section Review

Living Things Need Energy
USING KEY TERMS

1. Use each of the following terms in a separate sentence: *herbivores, carnivores,* and *omnivores.*

_____

_____

_____

2. In your own words, write a definition for each of the following terms: *food chain, food web,* and *energy pyramid.*

_____

_____

_____

UNDERSTANDING KEY IDEAS

_____ **3.** Herbivores, carnivores, and scavengers are all examples of
   a. producers.
   b. decomposers.
   c. consumers.
   d. omnivores.

4. Explain the importance of decomposers in an ecosystem.

_____

_____

5. Describe how producers, consumers, and decomposers are linked in a food chain.

_____

_____

6. Describe how energy flows through a food web.

_____

_____

_____

MATH SKILLS

7. The plants in each square meter of an ecosystem obtained 20,810 Calories of energy from sunlight per year. The herbivores in that ecosystem ate all the plants but obtained only 3,370 Calories of energy. How much energy did the plants use? Show your work below.

CRITICAL THINKING

8. Identifying Relationships Draw two food chains, and depict how they link together to form a food web.

9. Applying Concepts Are consumers found at the top or bottom of an energy pyramid? Explain your answer.

_____

_____

_____

_____

10. Predicting Consequences What would happen if a species disappeared from an ecosystem?

_____

_____

_____

_____

Skills Worksheet

Section Review

Types of Interactions

USING KEY TERMS

1. In your own words, write a definition for the term *carrying capacity*.

_____

_____

2. Use each of the following terms in a separate sentence: *mutualism, commensalism,* and *parasitism.*

_____

_____

_____

UNDERSTANDING KEY IDEAS

_____ **3.** Which of the following is NOT a prey adaptation?
   a. camouflage
   b. chemical defenses
   c. warning coloration
   d. parasitism

4. Identify two things organisms compete with one another for.

_____

_____

5. Briefly describe one example of a predator-prey relationship. Identify the predator and the prey.

_____

_____

_____

_____

Section Review *continued*

CRITICAL THINKING

6. Making Comparisons Compare coevolution with symbiosis.

_____

_____

_____

_____

7. Identifying Relationships Explain the probable relationship between the giant *Rafflesia* flower, which smells like rotting meat, and the carrion flies that buzz around it. (Hint: *Carrion* means "rotting flesh.")

_____

_____

_____

_____

_____

8. Predicting Consequences Predict what might happen if all of the ants were removed from an acacia tree.

_____

_____

_____

_____

_____

Name _____ Class _____ Date _____

INTERPRETING GRAPHICS

The population graph below shows the growth of a species of *Paramecium* (single-celled microorganism) over 18 days. Food was added to the test tube occasionally. Use this graph to answer the questions that follow.

Paramecium caudatum Growth

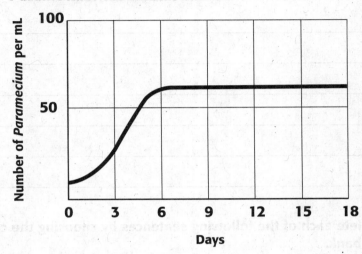

9. What is the carrying capacity of the test tube as long as food is added?

_____

_____

_____

10. Predict what will happen if no more food is added.

_____

_____

_____

11. What keeps the number of *Paramecium* at a steady level?

_____

_____

_____

Skills Worksheet

Chapter Review

USING KEY TERMS

1. Use each of the following terms in a separate sentence: *symbiosis, mutualism, commensalism,* and *parasitism.*

_____

_____

_____

_____

_____

_____

Complete each of the following sentences by choosing the correct term from the word bank.

biotic	abiotic	ecosystem	community

2. The environment includes _____ factors including water, rocks, and light.

3. The environment also includes _____, or living, factors.

4. A community of organisms and their environment is called a(n)

_____.

For each pair of terms, explain how the meanings of the terms differ.

5. *community* and *population*

_____

_____

_____

6. *ecosystem* and *biosphere*

_____

_____

_____

_____

Chapter Review *continued*

7. *producers* and *consumers*

_____

_____

_____

UNDERSTANDING KEY IDEAS
Multiple Choice

_____ **8.** A tick sucks blood from a dog. In this relationship, the tick is the
_____ and the dog is the _____.
   a. parasite, prey
   b. predator, host
   c. parasite, host
   d. host, parasite

_____ **9.** Resources such as water, food, or sunlight are likely to be limiting factors
   a. when population size is decreasing.
   b. when predators eat their prey.
   c. when the population is small.
   d. when a population is approaching the carrying capacity.

_____ **10.** Nature's recyclers are
   a. predators.
   b. decomposers.
   c. producers.
   d. omnivores.

_____ **11.** A beneficial association between coral and algae is an example of
   a. commensalism.
   b. parasitism.
   c. mutualism.
   d. predation.

_____ **12.** The process by which energy moves through an ecosystem can be represented by
   a. food chains.
   b. energy pyramids.
   c. food webs.
   d. All of the above

_____ **13.** Which organisms does the base of an energy pyramid represent?
   a. producers
   b. carnivores
   c. herbivores
   d. scavengers

_____ **14.** Which of the following is the correct order in a food chain?
   a. sun→producers→herbivores→scavengers→carnivores
   b. sun→consumers→predators→parasites→hosts
   c. sun→producers→decomposers→consumers→omnivores
   d. sun→producers→herbivores→carnivores→scavengers

_____15. Remoras and sharks have a relationship that is best described as
   a. mutualism.
   b. commensalism.
   c. predator and prey.
   d. parasitism.

Short Answer

16. Describe how energy flows through a food web.

_____

_____

_____

17. Explain how the food web changed when the gray wolf disappeared from Yellowstone National Park.

_____

_____

_____

18. How are the competition between two trees of the same species and the competition between two different species of trees similar?

_____

_____

_____

19. How do limiting factors affect the carrying capacity of an environment?

_____

_____

_____

20. What is coevolution?

_____

_____

_____

Chapter Review *continued*

CRITICAL THINKING

21. Concept Mapping Use the following terms to create a concept map: *herbivores, organisms, producers, populations, ecosystems, consumers, communities, carnivores,* and *biosphere.*

22. Identifying Relationships Could a balanced ecosystem contain producers and consumers but not decomposers? Why or why not?

_____

_____

23. Predicting Consequences Some biologists think that certain species, such as alligators and wolves, help maintain biological diversity in their ecosystems. Predict what might happen to other organisms, such as gar fish or herons, if alligators were to become extinct in the Florida Everglades.

_____

_____

24. Expressing Opinions Do you think there is a carrying capacity for humans? Why or why not?

_____

_____

Chapter Review *continued*

INTERPRETING GRAPHICS

Use the energy pyramid below to answer the questions that follow.

25. According to the energy pyramid, are there more prairie dogs or plants?

_____

_____

_____

26. What level has the most energy?

_____

_____

_____

27. Would an energy pyramid such as this one exist in nature?

_____

_____

_____

28. How could you change this pyramid to look like one representing a real ecosystem?

_____

_____

_____

_____

_____

Skills Worksheet

Section Review

The Cycles of Matter

USING KEY TERMS

For each pair of terms, explain how the meanings of the terms differ.

1. *evaporation* and *condensation*

_____

_____

2. *decomposition* and *combustion*

_____

_____

UNDERSTANDING KEY IDEAS

_____ **3.** Nitrogen fixation
    a. is done only by plants.
    b. is done mostly by bacteria.
    c. is how animals make proteins.
    d. is a form of decomposition.

4. Describe the water cycle.

_____

_____

5. Describe the carbon cycle.

_____

_____

MATH SKILLS

6. The average person in the United States uses about 78 gal of water each day. How many liters of water does this equal? How many liters of water will the average person use in a year? Show your work below.

CRITICAL THINKING

7. Analyzing Processes Draw a simple diagram of each of the cycles discussed in this section. Draw lines between the cycles to show how parts of each cycle are related.

8. Applying Concepts Give an example of how the calcium in an animal's bones might be cycled back into the environment.

_____

_____

_____

_____

Name _____ Class _____ Date _____

Skills Worksheet

Section Review

Ecological Succession

USING KEY TERMS

1. In your own words, write a definition for the term *succession*.

_____

_____

UNDERSTANDING KEY IDEAS

_____ **2.** An area where a glacier has just melted away will begin the process of
   a. primary succession.
   b. secondary succession.
   c. stability.
   d. regrowth.

3. Describe succession that takes place in an abandoned field.

_____

_____

_____

4. Describe a mature community. How does a mature community develop?

_____

_____

_____

MATH SKILLS

5. The fires in 1988 burned 739,000 of the 2.2 million acres that make up Yellowstone National Park. What percentage of the park was burned? Show your work below.

CRITICAL THINKING

6. Applying Concepts Give an example of a community that has a high degree of biodiversity, and an example of one that has a low degree of biodiversity.

_____

_____

_____

_____

7. Analyzing Ideas Explain why soil formation is always the first stage of primary succession. Does soil formation ever stop? Explain your answer.

_____

_____

_____

_____

Chapter Review

USING KEY TERMS

Complete each of the following sentences by choosing the correct term from the word bank.

evaporation	condensation	precipitation
decomposition	combustion	succession

1. The breakdown of dead materials into carbon dioxide and water is called

_____.

2. The gradual development of a community over time is called

_____.

3. During _____, the heat causes water to change from

liquid to vapor.

4. _____ is the process of burning a substance.

5. Water that falls from the atmosphere to the land and oceans is

_____.

6. In the process of _____, water vapor cools and returns to

a liquid state.

UNDERSTANDING KEY IDEAS

Multiple Choice

_____ **7.** Clouds form in the atmosphere through the process of
- **a.** precipitation.
- **b.** respiration.
- **c.** condensation.
- **d.** decomposition.

_____ **8.** Which of the following statements about groundwater is true?
- **a.** It stays underground for a few days.
- **b.** It is stored in underground caverns or porous rock.
- **c.** It is salty like ocean water.
- **d.** It never reenters the water cycle.

_____ **9.** Burning gas in an automobile is a type of
- **a.** combustion.
- **b.** respiration.
- **c.** decomposition.
- **d.** photosynthesis.

_____ **10.** Nitrogen in the form of a gas can be used directly by some kinds of
- **a.** plants.
- **b.** animals.
- **c.** bacteria.
- **d.** fungi.

Chapter Review *continued*

_____11. Bacteria are most important in the process of
    a. combustion.   **c.** nitrogen fixation.
    b. condensation.   **d.** evaporation.

_____12. The pioneer species on bare rock are usually
    a. ferns.   **c.** mosses.
    b. pine trees.   **d.** lichens.

_____13. Which of the following is an example of primary succession?
    a. the recovery of Yellowstone National Park following the fires of 1988
    b. the appearance of lichens and mosses in an area where a glacier has recently melted away
    c. the growth of weeds in a field after a farmer stops using the field
    d. the growth of weeds in an empty lot that is no longer being mowed

_____14. One of the most common plants in a recently abandoned farm field is
    a. oak or maple trees.
    b. pine trees.
    c. mosses.
    d. crabgrass.

Short Answer

15. List four places where water can go after it falls as precipitation.

_____

_____

_____

_____

16. In what forms can water on Earth be found?

_____

_____

_____

_____

17. What role do animals have in the carbon cycle?

_____

_____

_____

_____

18. What roles do humans have in the carbon cycle?

_____

_____

_____

19. Earth's atmosphere is mostly made up of what substance?

_____

_____

_____

20. Compare and contrast the two forms of succession.

_____

_____

_____

CRITICAL THINKING

21. Concept Mapping Use the following terms to create a concept map: *abandoned farmland, lichens, bare rock, soil formation, horseweed, succession, forest fire, primary succession, secondary succession,* and *pioneer species.*

22. Identifying Relationships Is snow a part of the water cycle? Why or why not?

_____

_____

_____

23. Analyzing Processes Make a list of several places where water might be found on Earth. For each item on your list, state how it is part of the water cycle.

_____

_____

_____

24. Forming Hypotheses Predict what would happen if the water on Earth suddenly stopped evaporating.

_____

_____

_____

25. Forming Hypotheses Predict what would happen if all of the bacteria on Earth suddenly disappeared.

_____

_____

_____

26. Making Inferences Describe why a lawn usually doesn't go through succession.

_____

_____

_____

27. Making Inferences Can one scientist observe all of the stages of secondary succession on an abandoned field? Explain your answer.

_____

_____

_____

INTERPRETING GRAPHICS

The graph below shows how water is used each day by an average household in the United States. Use the graph to answer the questions that follow.

Average Household Daily Water Use

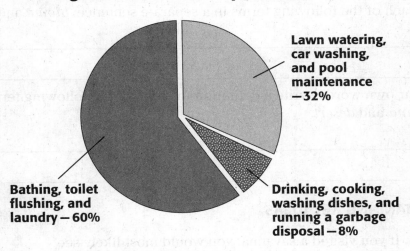

Lawn watering, car washing, and pool maintenance —32%

Bathing, toilet flushing, and laundry — 60%

Drinking, cooking, washing dishes, and running a garbage disposal — 8%

_____28. According to this graph, which of the following activities uses the greatest amount of water?
   a. bathing
   b. toilet flushing
   c. washing laundry
   d. There is not enough information to determine the answer.

29. An average family used 380 L of water per day, until they stopped washing their car, stopped watering their lawn, and stopped using their pool. Now, how much water per day do they use?

_____

_____

_____

Skills Worksheet

Section Review

Land Biomes

USING KEY TERMS

1. Use each of the following terms in a separate sentence: *biome* and *tundra*.

   _____

   _____

   _____

2. In your own words, write a definition for each of the following terms: *savanna* and *desert*.

   _____

   _____

   _____

UNDERSTANDING KEY IDEAS

_____ 3. If you visited a savanna, you would most likely see
   a. large herds of grazing animals, such as zebras, gazelles, and wildebeests.
   b. dense forests stretching from horizon to horizon.
   c. snow and ice throughout most of the year.
   d. trees that form a continuous green roof, called the *canopy*.

_____ 4. Components of a desert ecosystem include
   a. a hot, dry climate.
   b. plants that grow far apart.
   c. animals that are active mostly at night.
   d. All of the above

5. List seven land biomes that are found on Earth.

   _____

   _____

   _____

   _____

   _____

   _____

   _____

Name _____ Class _____ Date _____

Section Review *continued*

6. What are two things that characterize a biome?

_____

_____

_____

CRITICAL THINKING

7. Making Inferences While excavating an area in the desert, a scientist discovers the fossils of very large trees and ferns. What might the scientist conclude about biomes in this area?

_____

_____

_____

8. Analyzing Ideas Tundra receives very little rainfall. Could tundra accurately be called a *frozen desert?* Explain your answer.

_____

_____

_____

INTERPRETING GRAPHICS

Use the bar graph below to answer the questions that follow.

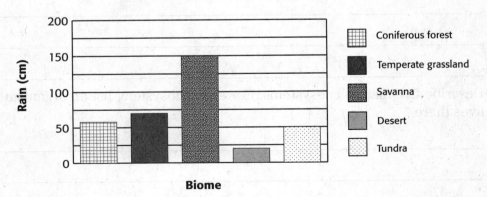

Rainfall on Biomes

9. Which biomes receive 50 cm or more of rain?

_____

10. Which biome receives the smallest amount of rain? the largest amount of rain?

_____

Skills Worksheet

Section Review

Marine Ecosystems

USING KEY TERMS

1. Use each of the following terms in a separate sentence: *plankton* and *estuary*.

_____

_____

UNDERSTANDING KEY IDEAS

_____ **2.** Water temperature
   a. has no effect on the animals in a marine ecosystem.
   b. affects the types of organisms that can live in a marine ecosystem.
   c. decreases gradually as water gets deeper.
   d. increases as water gets deeper.

3. What are three abiotic factors that affect marine ecosystems?

_____

_____

_____

4. Describe four major ocean zones.

_____

_____

_____

_____

5. Describe five marine ecosystems. For each ecosystem, list an organism that lives there.

_____

_____

_____

_____

_____

| Section Review *continued*

MATH SKILLS

6. The ocean covers about 71% of the Earth's surface. If the total surface area of the Earth is about 510 million square kilometers, how many square kilometers are covered by the ocean? Show your work below.

CRITICAL THINKING

7. Making Inferences Animals in the Sargasso Sea hide from predators by blending in with the sargassum. Color is only one way to blend in. What is another way that animals can blend in with sargassum?

_____

_____

_____

_____

8. Identifying Relationships Many fishes and other organisms that live in the deep ocean produce light. What are two ways in which this light might be useful?

_____

_____

_____

9. Applying Concepts Imagine that you are studying animals that live in intertidal zones. You just discovered a new animal. Describe the animal and adaptations the animal has to survive in the intertidal zone.

_____

_____

_____

_____

Section Review

Freshwater Ecosystems

USING KEY TERMS

1. Use the following terms in the same sentence: *wetland*, *marsh*, and *swamp*.

_____

_____

_____

UNDERSTANDING KEY IDEAS

_____ **2.** A major abiotic factor in freshwater ecosystems is the
   a. source of the water.
   b. speed of the water.
   c. width of the stream or river.
   d. None of the above

3. Describe the three zones of a lake.

_____

_____

_____

4. Explain how a lake can become a forest over time.

_____

_____

_____

_____

| Section Review *continued*

MATH SKILLS

5. Sunlight can penetrate a certain lake to a depth of 15 m. The lake is five and a half times deeper than the depth to which light can penetrate. In meters, how deep is the lake? Show your work below.

CRITICAL THINKING

6. Making Inferences When bacteria decompose material in a pond, the oxygen in the water may be used up. So, fishes in the pond die. How might the absence of fish lead to a pond filling faster?

_____

_____

_____

_____

7. Applying Concepts Imagine a steep, rocky stream. What kinds of adaptations might animals living in this stream have? Explain your answer.

_____

_____

_____

_____

Skills Worksheet

Chapter Review

USING KEY TERMS

1. In your own words, write a definition for the following terms: *biome* and *tundra*.

_____

_____

2. Use each of the following terms in a separate sentence: *intertidal zone*, *neritic zone*, and *oceanic zone*.

_____

_____

_____

For each pair of terms, explain how the meanings of the terms differ.

3. *savanna* and *desert*

_____

_____

4. *open-water zone* and *deep-water zone*

_____

_____

5. *marsh* and *swamp*

_____

_____

UNDERSTANDING KEY IDEAS

Multiple Choice

_____ **6.** Trees that lose their leaves in the winter are called

    a. evergreen trees.   **c.** deciduous trees.

    b. coniferous trees.   **d.** None of the above

_____ **7.** In which major ocean zone are plants and animals exposed to air for part of the day?

    a. intertidal zone   **c.** oceanic zone

    b. neritic zone   **d.** benthic zone

_____ **8.** An abiotic factor that affects marine ecosystems is
     a. the temperature of the water.
     b. the depth of the water.
     c. the amount of sunlight that passes through the water.
     d. All of the above

_____ **9.** _____ is a marine ecosystem that includes mudflats, sandy beaches, and rocky shores.
     a. An intertidal area
     b. Polar ice
     c. A coral reef
     d. The Sargasso Sea

Short Answer

10. What are seven land biomes?

_____

_____

_____

_____

11. Explain how a small lake can become a forest.

_____

_____

_____

_____

12. What are two factors that characterize biomes?

_____

_____

13. Describe the three zones of a lake.

_____

_____

_____

14. How do rivers form?

_____

_____

15. What are three abiotic factors in land biomes? three abiotic factors in marine ecosystems? an abiotic factor in freshwater ecosystems?

_____

_____

_____

CRITICAL THINKING

16. Concept Mapping Use the following terms to create a concept map: *plants and animals, tropical rain forest, tundra, biomes, permafrost, canopy, desert,* and *abiotic factors.*

Chapter Review *continued*

17. Making Inferences Plankton use photosynthesis to make their own food. They need sunlight for photosynthesis. Which of the four major ocean zones can support plankton growth? Explain your answer.

_____

_____

_____

_____

18. Predicting Consequences Wetlands, such as marshes and swamps, play an important role in flood control. Wetlands also help replenish underground water supplies. Predict what might happen if a wetland dries out.

_____

_____

_____

_____

19. Analyzing Ideas A scientist has a new hypothesis. He or she thinks that savannas and deserts are part of one biome rather than two separate biomes. Based on what you've learned, decide if the scientist's hypothesis is correct. Explain your answer.

_____

_____

_____

_____

20. Applying Concepts Imagine that you are a scientist. You are studying an area that gets about 100 cm of rain each year. The average summer temperatures are near 30°C. What biome are you in? What are some plants and animals you will likely encounter? If you stayed in this area for the winter, what kind of preparations might you need to make?

_____

_____

_____

_____

| Chapter Review *continued*

INTERPRETING GRAPHICS
Use the graphs below to answer the questions that follow.

Average Monthly Precipitation

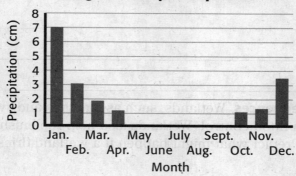

Average Monthly High Temperatures

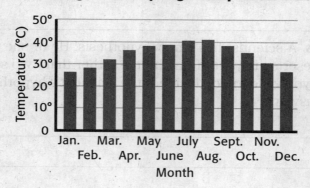

21. Which biome is most likely found in the region described by the graphs above? Explain your answer.

_____

_____

_____

22. How many centimeters of rain fell in the region during the course of the year?

_____

_____

_____

23. Which month is the hottest in the region? the coolest in the region?

_____

_____

24. What is the average monthly precipitation for the month that has the highest average high temperature?

_____

_____

Section Review

Environmental Problems

USING KEY TERMS

The statements below are false. For each statement, replace the underlined term to make a true statement.

1. Coal is a <u>renewable resource</u>.

_____

2. <u>Overpopulation</u> is the number and variety of organisms in an area.

_____

UNDERSTANDING KEY IDEAS

_____ **3.** Which of the following can cause pollution?
   a. noise
   b. garbage
   c. chemicals
   d. All of the above

_____ **4.** Pollution
   a. does not affect humans.
   b. can make humans sick.
   c. makes humans sick only after many years.
   d. None of the above

5. Compare renewable and nonrenewable resources.

_____

_____

_____

_____

6. Why has human population growth increased?

_____

_____

_____

Section Review *continued*

7. What is an exotic species?

_____

_____

8. How does habitat destruction affect biodiversity?

_____

_____

MATH SKILLS

9. Jodi's family produces 48 kg of garbage each week. What is the percentage decrease if they reduce the amount of garbage to 40 kg per week? Show your work below.

CRITICAL THINKING

10. Applying Concepts Explain how each of the following can help people but harm the environment: hospitals, old refrigerators, and road construction.

_____

_____

_____

11. Making Inferences Explain how human population growth is related to pollution problems.

_____

_____

_____

12. Predicting Consequences How can the pollution of marine habitats affect humans?

_____

_____

_____

Skills Worksheet

Section Review

Environmental Solutions

USING KEY TERMS

1. Use each of the following terms in a separate sentence: *conservation* and *recycling*.

_____

_____

UNDERSTANDING KEY IDEAS

_____ 2. Which of the following is NOT a strategy to protect the environment?
   a. preserving entire habitats
   b. using pesticides that target all insects
   c. reducing deforestation
   d. increasing the use of solar power

_____ 3. Conservation
   a. has little effect on the environment.
   b. is the use of more natural resources.
   c. involves using more fossil fuels.
   d. can prevent pollution.

4. Describe the three Rs.

_____

_____

_____

5. Describe why biodiversity is important. How can biodiversity be protected?

_____

_____

CRITICAL THINKING

6. **Applying Concepts** Liza rode her bike to the store. She bought items that had little packaging and put her purchases into her backpack. Describe how Liza practiced conservation.

_____

_____

_____

Section Review *continued*

7. Identifying Relationships How does conservation of resources also reduce pollution and protect habitats?

_____

_____

_____

INTERPRETING GRAPHICS

Use the pie graph below to answer the questions that follow.

Land Use in the United States

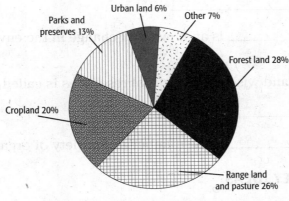

Source: Natural Resources Conservation Service.

8. If half of the forest land were made into preserves, what percentage of total land would be parks and preserves?

_____

_____

_____

9. If 10% of the cropland were not planted, what percentage of land would be used for crops?

_____

_____

_____

Skills Worksheet)

Chapter Review

USING KEY TERMS

Complete each of the following sentences by choosing the correct term from the word bank.

conservation   pollution   nonrenewable resource
biodiversity   overpopulation   renewable resource
recycling

1. A(n) _____ is a resource that is replaced at a much

slower rate than it is used.

2. The presence of too many individuals in a population for available resources

is called _____.

3. _____ is an unwanted change in the environment caused

by wastes.

4. The preservation and wise use of natural resources is called

_____.

5. _____ is the number and variety of organisms in an area.

UNDERSTANDING KEY IDEAS

Multiple Choice

_____ **6.** Preventing habitat destruction is important because
     a. organisms do not live independently of each other.
     b. protection of habitats is a way to promote biodiversity.
     c. the balance of nature could be disrupted if habitats were destroyed.
     d. All of the above

_____ **7.** Exotic species
     a. do not affect native species.
     b. are species that make a home for themselves in a new place.
     c. are not introduced by human activity.
     d. do not take over an area.

_____ **8.** A renewable resource
     a. is a natural resource that can be replaced as quickly as it is used.
     b. is a natural resource that takes thousands or millions of years to be
        replaced.
     c. includes fossil fuels, such as coal or oil.
     d. will eventually run out.

Name _____ Class _____ Date _____

Chapter Review *continued*

Short Answer

9. Describe how you can use the three Rs to conserve resources.

_____

_____

_____

_____

10. What are five kinds of pollutants?

_____

_____

_____

_____

11. Explain why human population growth has increased.

_____

_____

_____

_____

12. What are two things that can be done to maintain biodiversity?

_____

_____

_____

13. List five environmental strategies.

_____

_____

_____

_____

Chapter Review *continued*

CRITICAL THINKING

14. Concept Mapping Use the following terms to create a concept map: *pollution, radioactive wastes, gases, pollutants, CFCs, PCBs, hazardous wastes, chemicals, noise,* and *garbage.*

Chapter Review *continued*

15. Analyzing Ideas How might deforestation have contributed to the extinction of some species?

_____

_____

_____

16. Predicting Consequences Imagine that the supply of fossil fuels is going to run out in 50 years. What will happen if people are not prepared when the supply runs out? What might be done to prepare for such an event?

_____

_____

_____

_____

17. Evaluating Conclusions A scientist thinks that farms should be planted with many different kinds of crops instead of a single crop. Based on what you learned about biodiversity, evaluate the scientist's conclusion. What problems might this cause?

_____

_____

_____

18. Applying Concepts Imagine that a new species has moved into a local habitat. The species feeds on some of the same plants that the native species do, but it has no natural predators. Describe what might happen to local habitats as a result.

_____

_____

_____

Chapter Review *continued*

19. Making Inferences Many scientists think that forests are nonrenewable resources. Explain why they might have this opinion.

_____

_____

_____

_____

INTERPRETING GRAPHICS

The line graph below shows the concentration of carbon dioxide in the atmosphere between 1958 and 1994. Use this graph to answer the questions that follow.

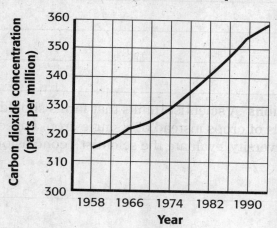

Carbon Dioxide in the Atmosphere

20. What was the concentration of carbon dioxide in parts per million in 1960? in 1994?

_____

_____

21. What is the average change in carbon dioxide concentration every 4 years?

_____

_____

22. If the concentration of carbon dioxide continues to change at the rate shown in the graph, what will the concentration be in 2010?

_____

_____

Skills Worksheet

Section Review

Body Organization
USING KEY TERMS

1. Use the following terms in the same sentence: *homeostasis*, *tissue*, and *organ*.

_____

_____

_____

UNDERSTANDING KEY IDEAS

_____ **2.** Which of the following statements describes how tissues, organs, and organ systems are related?
  a. Organs form tissues, which form organ systems.
  b. Organ systems form organs, which form tissues.
  c. Tissues form organs, which form organ systems.
  d. None of the above

3. List the 11 organ systems.

_____

_____

_____

_____

_____

_____

_____

_____

_____

Section Review continued

MATH SKILLS

4. The human skeleton has 206 bones. The human skull has 22 bones. What percentage of human bones are skull bones? Show your work below.

CRITICAL THINKING

5. Applying Concepts Tanya went to a restaurant and ate a hamburger. Describe how Tanya used five organ systems to eat and digest her hamburger.

_____

_____

_____

_____

_____

_____

6. Predicting Consequences Predict what might happen if the human body did not have specialized cells, tissues, organs, and organ systems to maintain homeostasis.

_____

_____

_____

_____

_____

Skills Worksheet

Section Review

The Skeletal System
USING KEY TERMS

1. In your own words, write a definition for the term *skeletal system*.

   _____

   _____

   _____

UNDERSTANDING KEY IDEAS

_____ 2. Which of the following is NOT an organ of the skeletal system?
   a. bone
   b. cartilage
   c. muscle
   d. None of the above

3. Describe four functions of bones.

   _____

   _____

   _____

   _____

4. What are three joints?

   _____

   _____

   _____

5. Describe two diseases that affect the skeletal system.

   _____

   _____

   _____

   _____

   _____

Section Review *continued*

MATH SKILLS

6. A broken bone usually heals in about six weeks. A mild sprain takes one-third as long to heal. In days, about how long does it take a mild sprain to heal? Show your work below.

CRITICAL THINKING

7. Identifying Relationships Red bone marrow produces blood cells. Children have red bone marrow in their long bones, while adults have yellow bone marrow, which stores fat. Why might adults and children have different kinds of marrow?

_____

_____

_____

_____

_____

8. Predicting Consequences What might happen if children's bones didn't have growth plates of cartilage?

_____

_____

_____

_____

_____

Skills Worksheet

Section Review

The Muscular System

USING KEY TERMS

1. In your own words, write a definition for the term *muscular system*.

_____

_____

UNDERSTANDING KEY IDEAS

_____ **2.** Muscles
   a. work in pairs.
   b. move bones by relaxing.
   c. get smaller when exercised.
   d. All of the above

3. Describe three kinds of muscle.

_____

_____

_____

4. List two kinds of exercise. Give an example of each.

_____

_____

5. Describe two muscular system injuries.

_____

_____

_____

_____

MATH SKILLS

6. If Trey can do one curl-up every 2.5 s, about how long will it take him to do 35 curl-ups? Show your work below.

CRITICAL THINKING

7. Applying Concepts Describe some of the muscle action needed to pick up a book. Include flexors and extensors in your description.

_____

_____

_____

_____

_____

8. Predicting Consequences If aerobic exercise improves heart strength, what likely happens to heart rate as the heart gets stronger? Explain your answer.

_____

_____

_____

_____

_____

_____

Skills Worksheet

Section Review

The Integumentary System

USING KEY TERMS

1. In your own words, write a definition for each of the following terms: *integumentary system*, *epidermis*, and *dermis*.

_____

_____

_____

_____

UNDERSTANDING KEY IDEAS

_____ 2. Which of the following is NOT a function of skin?
   a. to regulate body temperature
   b. to keep water in the body
   c. to move your body
   d. to get rid of wastes

3. Describe the two layers of skin.

_____

_____

_____

4. How do hair and nails develop?

_____

_____

5. Describe how a cut heals.

_____

_____

_____

| Section Review *continued*

MATH SKILLS

6. On average, hair grows 0.3 mm per day. How many millimeters does hair grow in 30 days? in a year? Show your work below.

CRITICAL THINKING

7. Making Inferences Why do you feel pain when you pull on your hair or nails, but not when you cut them?

_____

_____

_____

_____

8. Analyzing Ideas The epidermis on the palms of your hands and on the soles of your feet is thicker than it is anywhere else on your body. Why might this skin need to be thicker?

_____

_____

_____

_____

_____

Skills Worksheet

Chapter Review

USING KEY TERMS

Complete each of the following sentences by choosing the correct term from the word bank.

homeostasis	organ	joint
skeletal system	tissue	muscular system
epidermis	dermis	integumentary system

1. A(n) _____ is a place where two or more bones meet.

2. _____ is the maintenance of a stable internal

environment.

3. The outermost layer of skin is the _____.

4. The organ system that includes skin, hair, and nails is the

_____.

5. A(n) _____ is made up of two or more tissues working

together.

6. The _____ supports and protects the body, stores minerals,

and allows movement.

UNDERSTANDING KEY IDEAS

Multiple Choice

_____ **7.** Which of the following lists shows the way in which the body is
organized?
  a. cells, organs, organ systems, tissues
  b. tissues, cells, organs, organ systems
  c. cells, tissues, organs, organ systems
  d. cells, tissues, organ systems, organs

_____ **8.** Which muscle tissue can be both voluntary and involuntary?
  a. smooth muscle   **c.** skeletal muscle
  b. cardiac muscle   **d.** All of the above

_____ **9.** The integumentary system
  a. helps regulate body temperature.
  b. helps the body move.
  c. stores minerals.
  d. None of the above

Chapter Review *continued*

_____ **10.** Muscles
      a. work in pairs.
      b. can be voluntary or involuntary.
      c. become stronger if exercised.
      d. All of the above

Short Answer

11. How do muscles move bones?

_____

_____

_____

12. Describe the skeletal system, and list four functions of bones.

_____

_____

_____

_____

13. Give an example of how organ systems work together.

_____

_____

_____

14. List three injuries and two diseases that affect the skeletal system.

_____

_____

_____

15. Compare aerobic exercise and resistance exercise.

_____

_____

_____

16. What are two kinds of damage that may affect skin?

_____

_____

_____

CRITICAL THINKING

17. Concept Mapping Use the following terms to create a concept map:
  tissues, muscle tissue, connective tissue, cells, organ systems, organs, epithelial tissue, and *nervous tissue*.

| Chapter Review *continued*

18. **Making Comparisons** Compare the shapes of the bones of the human skull with the shapes of the bones of the human leg. How do the shapes differ? Why are the shapes important?

_____

_____

_____

_____

19. **Making Inferences** Compare your elbows and fingertips in terms of the texture and sensitivity of the skin on these parts of your body. Why might the skin on these body parts differ?

_____

_____

_____

_____

20. **Making Inferences** Imagine that you are building a robot. Your robot will have a skeleton similar to a human skeleton. If the robot needs to be able to move a limb in all directions, what kind of joint would be needed? Explain your answer.

_____

_____

_____

21. **Analyzing Ideas** Human bones are dense and are often filled with marrow. But many bones of birds are hollow. Why might birds have hollow bones?

_____

_____

_____

22. **Identifying Relationships** Why might some muscles fail to work properly if a bone is broken?

_____

_____

_____

Chapter Review *continued*

INTERPRETING GRAPHICS

Use the cross section of skin below to answer the questions that follow.

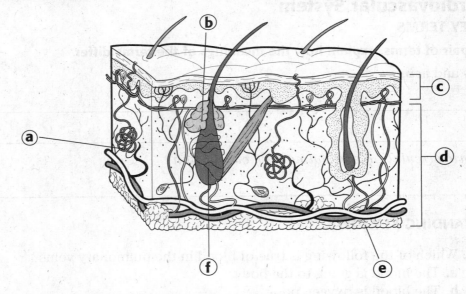

23. What is d called? What substance is most abundant in this layer?

_____

24. What is the name and function of a?

_____

25. What is the name and function of b?

_____

26. Which letter corresponds to the part of the skin that is made up of epithelial tissue that contains dead cells?

_____

27. Which letter corresponds to the part of the skin from which hair grows? What is this part called?

_____

_____

Skills Worksheet

Section Review

The Cardiovascular System

USING KEY TERMS

For each pair of terms, explain how the meanings of the terms differ.

1. *artery* and *vein*

   _____

   _____

2. *systemic circulation* and *pulmonary circulation*

   _____

UNDERSTANDING KEY IDEAS

_____ 3. Which of the following is true of blood in the pulmonary veins?
   a. The blood is going to the body.
   b. The blood is oxygen poor.
   c. The blood is going to the lungs.
   d. The blood is oxygen rich.

4. What are the five parts of the cardiovascular system? Describe the functions of each part.

   _____

   _____

   _____

   _____

5. What is the difference between a heart attack and heart failure?

   _____

   _____

MATH SKILLS

6. An adult male's heart pumps about 2.8 million liters of blood a year. If his heart beats 70 times a minute, how much blood does his heart pump with each beat? Show your work below.

CRITICAL THINKING

7. Identifying Relationships How is the structure of capillaries related to their function?

_____

_____

_____

8. Making Inferences One of aspirin's effects is that it prevents platelets from being too "sticky." Why might doctors prescribe aspirin for patients who have had a heart attack?

_____

_____

_____

9. Analyzing Ideas Veins and arteries are everywhere in your body. When a pulse is taken, it is usually taken at an artery in the neck or wrist. Explain why.

_____

_____

10. Making Comparisons Why is the structure of arteries different from the structure of capillaries?

_____

_____

Skills Worksheet

Section Review

Blood

USING KEY TERMS

1. Use each of the following terms in a separate sentence: *blood* and *blood pressure*.

_____

_____

UNDERSTANDING KEY IDEAS

_____ 2. A person with type B blood can donate blood to people with which type(s) of blood?
   a. B, AB
   b. A, AB
   c. AB only
   d. All types

3. List the four main components of blood and tell what each component does.

_____

_____

_____

_____

4. Why is it important for a doctor to know a patient's blood type?

_____

_____

MATH SKILLS

5. A person has a systolic pressure of 174 mm Hg. What percentage of normal (120 mm Hg) is this? Show your work below.

CRITICAL THINKING

6. Identifying Relationships How does the body use blood and blood vessels to help maintain proper body temperature?

_____

_____

_____

_____

_____

_____

7. Predicting Consequences Some blood conditions and diseases affect the ability of red blood cells to deliver oxygen to cells of the body. Predict what might happen to a person with a disease of that type.

_____

_____

_____

_____

_____

_____

Skills Worksheet

Section Review

The Lymphatic System
USING KEY TERMS

1. Use each of the following terms in a separate sentence: *lymph nodes*, *spleen*, and *tonsils*.

_____

_____

_____

UNDERSTANDING KEY IDEAS

_____ **2.** Lymph
  a. is the same as blood.
  b. is fluid in the cells.
  c. drains into your muscles.
  d. is fluid collected by lymphatic vessels.

3. Name six parts of the lymphatic system. Tell what each part does.

_____

_____

_____

_____

_____

_____

4. How are your cardiovascular and lymphatic systems related?

_____

_____

_____

_____

_____

MATH SKILLS

5. One cubic millimeter of blood contains 5 million RBCs and 10,000 WBCs. How many times more RBCs are there than WBCs? Show your work below.

CRITICAL THINKING

6. Expressing Opinions Some people have frequent, severe tonsil infections. These infections can be treated with medicine, and the infections usually go away after a few days. Do you think removing tonsils in such a case is a good idea? Explain.

_____

_____

_____

_____

_____

_____

7. Analyzing Ideas Why is it important that lymphatic tissue is spread throughout the body?

_____

_____

_____

_____

_____

_____

_____

Skills Worksheet

Section Review

The Respiratory System

USING KEY TERMS

For each pair of terms, explain how the meanings of the terms differ.

1. *pharynx* and *larynx*

_____

_____

UNDERSTANDING KEY IDEAS

_____ **2.** Which of the following are respiratory disorders?
   a. SARS, alveoli, and asthma
   b. alveoli, emphysema, and SARS
   c. larynx, asthma, and SARS
   d. SARS, emphysema, and asthma

3. Explain how breathing happens.

_____

_____

4. Describe how your cardiovascular and respiratory systems work together.

_____

_____

MATH SKILLS

5. Total lung capacity (TLC) is about 6 L. A person can exhale about 3.6 L. What percentage of TLC cannot be exhaled? Show your work below.

Section Review *continued*

CRITICAL THINKING

6. Interpreting Statistics About 6.3 million children in the United States have asthma. About 4 million of them had an asthma attack last year. What do these statistics tell you about the relationship between asthma and asthma attacks?

_____

_____

_____

_____

_____

_____

7. Identifying Relationships If a respiratory disorder causes lungs to fill with fluid, how might this affect a person's health?

_____

_____

_____

_____

_____

Skills Worksheet

Chapter Review

USING KEY TERMS

Complete each of the following sentences by choosing the correct term from the word bank.

red blood cells	veins	white blood cells
arteries	lymphatic system	larynx
alveoli	bronchi	respiratory system
trachea		

1. _____ deliver oxygen to the cells of the body.

2. _____ carry blood away from the heart.

3. The _____ helps the body fight pathogens.

4. The _____ contains the vocal cords.

5. The pathway of air through the respiratory system ends at the tiny sacs

called _____.

UNDERSTANDING KEY IDEAS

Multiple Choice

_____ **6.** Blood from the lungs enters the heart at the
    a. left ventricle.   **c.** right atrium.
    b. left atrium.   **d.** right ventricle.

_____ **7.** Blood cells are made
    a. in the heart.   **c.** from lymph.
    b. from plasma   **d.** in the bones.

_____ **8.** Which of the following activities is a function of the lymphatic system?
    a. returning excess fluid to the circulatory system
    b. delivering nutrients to the cells
    c. bringing oxygen to the blood
    d. pumping blood to all parts of the body

_____ **9.** Alveoli are surrounded by
    a. veins.   **c.** capillaries.
    b. muscles.   **d.** lymph nodes.

Chapter Review *continued*

_____10. What prevents blood from flowing backward in veins?
   a. platelets   **c.** muscles
   b. valves   **d.** cartilage

_____11. Air moves into the lungs when the diaphragm muscle
   a. contracts and moves down.   **c.** relaxes and moves down.
   b. contracts and moves up   **d.** relaxes and moves up.

Short Answer

12. What is the difference between pulmonary circulation and systemic circulation in the cardiovascular system?

_____

_____

13. Walton's blood pressure is 110/65. What do the two numbers mean?

_____

_____

14. What body process produces the carbon dioxide you exhale?

_____

_____

15. Describe how the cardiovascular system and the lymphatic system work together to keep your body healthy.

_____

_____

16. How is the spleen important to both the lymphatic system and the cardiovascular system?

_____

_____

17. Briefly describe the path that oxygen follows in your respiratory system and your cardiovascular system.

_____

_____

CRITICAL THINKING

18. Concept Mapping Use the following terms to create a concept map:
blood, oxygen, alveoli, capillaries, and carbon dioxide.

Name _____ Class _____ Date _____

Chapter Review *continued*

19. **Making Comparisons** Compare and contrast the functions of the cardiovascular system and the lymphatic system.

_____

_____

_____

20. **Identifying Relationships** Why do you think there are hairs in your nose?

_____

_____

_____

_____

21. **Applying Concepts** After a person donates blood, the blood is stored in
one-pint bags until it is needed for a transfusion. A healthy person has about 5
million RBCs in each cubic millimeter (1 mm^3) of blood.

   a. How many RBCs are in 1 mL of blood? (One milliliter is equal to 1 cm^3 and
   to 1,000 mm^3.)

   _____

   b. How many RBCs are there in 1 pt? (One pint is equal to 473 mL.)

   _____

22. **Predicting Consequences** What would happen if all of the red blood cells in
your blood disappeared?

_____

_____

_____

23. **Identifying Relationships** When a person is not feeling well, a doctor may
examine samples of the person's blood to see how many white blood cells are
present. Why would this information be useful?

_____

_____

_____

Chapter Review *continued*

INTERPRETING GRAPHICS

The diagram below shows how the human heart would look in cross section. Use the diagram to answer the questions that follow.

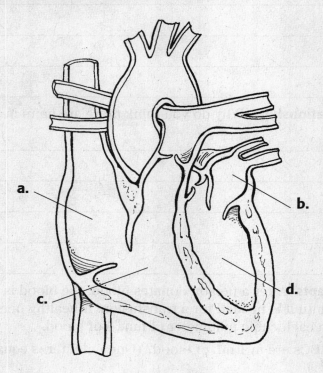

24. Which letter identifies the chamber that receives blood from systemic circulation? What is this chamber's name?

_____

25. Which letter identifies the chamber that receives blood from the lungs? What is this chamber's name?

_____

26. Which letter identifies the chamber that pumps blood to the lungs? What is this chamber's name?

_____

Skills Worksheet

Section Review

The Digestive System

USING KEY TERMS

1. Use each of the following terms in a separate sentence: *digestive system*, *large intestine*, and *small intestine*.

_____

_____

_____

UNDERSTANDING KEY IDEAS

_____ **2.** Which of the following is NOT a function of the liver?
    a. to secrete bile   **c.** to detoxify chemicals
    b. to store nutrients   **d.** to compact wastes

3. What is the difference between mechanical digestion and chemical digestion?

_____

_____

4. What happens to the food that you eat when it gets to your stomach?

_____

_____

5. Describe the role of the liver, gallbladder, and pancreas in digestion.

_____

_____

6. Put the following steps of digestion in order.

_____ **a.** Food is chewed by the teeth in the mouth.

_____ **b.** Water is absorbed by the large intestine.

_____ **c.** Food is reduced to chyme in the stomach.

_____ **d.** Food moves down the esophagus.

_____ **e.** Nutrients are absorbed by the small intestine.

_____ **f.** The pancreas releases enzymes.

Name _____ Class _____ Date _____

Section Review *continued*

CRITICAL THINKING

7. Evaluating Conclusions Explain the following statement: "Digestion begins in the mouth."

_____

_____

_____

_____

8. Identifying Relationships How would the inability to make saliva affect digestion?

_____

_____

_____

INTERPRETING GRAPHICS

9. Label and describe the function of each of the organs in the diagram below.

a. _____

_____

b. _____

_____

c. _____

_____

d. _____

_____

e. _____

_____

f. _____

_____

g. _____

_____

Skills Worksheet

Section Review

The Urinary System

USING KEY TERMS

1. In your own words, write a definition for the term *urinary system*.

_____

_____

UNDERSTANDING KEY IDEAS

_____ **2.** Which event happens first?
   a. Water is absorbed into blood.
   b. A large artery brings blood into the kidneys.
   c. Water enters the nephrons.
   d. The nephron separates water from wastes.

3. How do kidneys filter blood?

_____

_____

4. Describe three disorders of the urinary system.

_____

_____

_____

MATH SKILLS

5. A study has shown that 75% of teenage boys drink 34 oz of soda per day. How many 12-oz cans of soda would a boy drink in a week if he drank 34 oz per day? Show your work below.

CRITICAL THINKING

6. Applying Concepts Which of the following contains more water: the blood going into the kidney or the blood leaving it?

_____

_____

_____

_____

7. Predicting Consequences When people have one kidney removed, their other kidney can often keep their blood clean. But the remaining kidney often changes. Predict how the remaining kidney may change to do the work of two kidneys.

_____

_____

_____

_____

Skills Worksheet

Chapter Review

USING KEY TERMS

Complete each of the following sentences by choosing the correct term from the word bank.

pancreas	digestive system	large intestine
stomach	kidney	small intestine
nephron	urinary system	

1. The _____ secretes juices into the small intestine.

2. The saclike organ at the end of the esophagus is called the

_____.

3. The _____ is an organ that contains millions of nephrons.

4. A group of organs that removes waste from the blood and excretes it from the

body is called the _____.

5. The _____ is a group of organs that work together to

break down food.

6. Indigestible material is formed into feces in the _____.

UNDERSTANDING KEY IDEAS

Multiple Choice

_____ **7.** The hormone that signals the kidneys to make less urine is
    a. urea.
    b. caffeine.
    c. ADH.
    d. ATP.

_____ **8.** Which of the following organs aids digestion by producing bile?
    a. stomach
    b. pancreas
    c. small intestine
    d. liver

_____ **9.** The part of the kidney that filters the blood is the
    a. artery.
    b. ureter.
    c. nephron.
    d. urethra.

_____ **10.** The fingerlike projections that line the small intestine are called
    a. emulsifiers.
    b. fats.
    c. amino acids.
    d. villi.

_____ **11.** Which of the following is NOT part of the digestive tract?
    a. mouth
    b. kidney
    c. stomach
    d. rectum

Chapter Review *continued*

_____12. The soupy mixture of food, enzymes, and acids in the stomach is called
     a. chyme.   **c.** urea.
     b. villi.   **d.** vitamins.

_____13. The stomach helps with
     a. storing food.
     b. chemical digestion.   **c.** physical digestion.
                           d. All of the above

_____14. The gall bladder stores
     a. food.   **c.** bile.
     b. urine.   **d.** villi.

_____15. The esophagus connects the
     a. pharynx to the stomach.
     b. stomach to the small intestine.
     c. kidneys to the nephrons.
     d. stomach to the large intestine.

Short Answer

16. Why is it important for the pancreas to release bicarbonate into the small intestine?

_____

_____

_____

17. How does the structure of the small intestine help the small intestine absorb nutrients?

_____

_____

_____

18. What is a kidney stone?

_____

_____

_____

CRITICAL THINKING

19. Concept Mapping Use the following terms to create a concept map: *teeth, stomach, digestion, bile, saliva, mechanical digestion, gallbladder,* and *chemical digestion.*

20. Predicting Consequences How would digestion be affected if the liver were damaged?

_____

_____

_____

_____

21. Analyzing Processes When you put a piece of carbohydrate-rich food, such as bread, a potato, or a cracker, into your mouth, the food tastes bland. But if this food sits on your tongue for a while, the food will begin to taste sweet. What digestive process causes this change in taste?

_____

_____

_____

_____

22. Making Comparisons The recycling process for one kind of plastic begins with breaking the plastic into small pieces. Next, chemicals are used to break the small pieces of plastic down to its building blocks. Then, those building blocks are used to make new plastic. How is this process both like and unlike human digestion?

_____

_____

_____

_____

Chapter Review *continued*

INTERPRETING GRAPHICS

The bar graph below shows how long the average meal spends in each portion of your digestive tract. Use the graph below to answer the questions that follow.

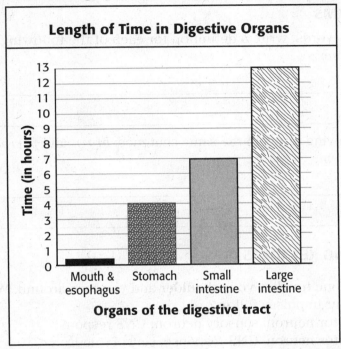

Length of Time in Digestive Organs

23. In which part of your digestive tract does the food spend the longest amount of time?

_____

_____

24. On average, how much longer does food stay in the small intestine than in the stomach?

_____

_____

25. Which organ mixes food with special substances to make chyme? Approximately how long does food remain in this organ?

_____

_____

26. Bile breaks large fat droplets into very small droplets. How long is the food in your body before it comes into contact with bile?

_____

_____

Name _____ Class _____ Date _____

Section Review

The Nervous System

USING KEY TERMS

1. In your own words, write a definition for each of the following terms: *neuron* and *nerve*.

_____

_____

2. Use the following terms in the same sentence: *brain* and *peripheral nervous system*.

_____

_____

UNDERSTANDING KEY IDEAS

_____ **3.** Someone touches your shoulder and you turn around. Which sequence do your impulses follow?
  a. motor neuron, sensory neuron, CNS response
  b. motor neuron, CNS response, sensory neuron
  c. sensory neuron, motor neuron, CNS response
  d. sensory neuron, CNS response, motor neuron

4. Describe one function of each part of the brain.

_____

_____

5. Compare the somatic nervous system with the autonomic nervous system.

_____

_____

6. Explain how a severe injury to the spinal cord can affect other parts of the body.

_____

_____

Section Review *continued*

CRITICAL THINKING

7. Applying Concepts Some medications slow a person's nervous system. These drugs are often labeled "May cause drowsiness." Explain why a person needs to know about this side effect.

_____

_____

_____

_____

8. Predicting Consequences Explain how your life would change if your autonomic nervous system suddenly stopped working.

_____

_____

_____

_____

INTERPRETING GRAPHICS

Use the drawing shown in your textbook for this Section Review to answer the questions that follow.

9. Which hemisphere of the brain recognizes and processes words, numbers, and letters? faces, places, and objects?

_____

_____

10. For a person whose left hemisphere is primarily in control, would it be easier to learn to play a new computer game by reading the rules and following instructions or by watching a friend play and imitating his actions?

_____

_____

Skills Worksheet

Section Review

Responding to the Environment
USING KEY TERMS

1. In your own words, write a definition for each of the following terms: *reflex* and *feedback mechanism*.

   _____

   _____

2. Use each of the following terms in a separate sentence: *retina* and *cochlea*.

   _____

   _____

   _____

UNDERSTANDING KEY IDEAS

_____ 3. Three sensations that receptors in the skin detect are
   a. light, smell, and sound.
   b. touch, pain, and odors.
   c. temperature, pressure, and pain.
   d. pressure, sound, and touch.

4. Explain how light and sight are related.

   _____

   _____

5. Describe how your senses of hearing, taste, and smell work.

   _____

   _____

6. Explain why you might have trouble seeing bright colors at a candlelit dinner.

   _____

   _____

7. How is your sense of taste similar to your sense of smell, and how do these senses work together?

   _____

   _____

Section Review *continued*

8. Describe how the feedback mechanism that regulates body temperature works.

_____

_____

MATH SKILLS

9. Suppose a nerve impulse must travel 0.90 m from your toe to your central
nervous system. If the impulse travels at 150 m/s, calculate how long it will
take the impulse to arrive. If the impulse travels at 0.2 m/s, how long will it
take the impulse to arrive? Show your work below.

CRITICAL THINKING

10. Making Inferences Why is it important for the human body to have reflexes?

_____

_____

_____

_____

11. Applying Concepts Rods help you detect objects and shapes in dim light.
Explain why it is important for human eyes to have both rods and cones.

_____

_____

_____

_____

Skills Worksheet

Section Review

The Endocrine System

USING KEY TERMS

1. Use the following terms in the same sentence: *endocrine system*, *glands*, and *hormone*.

   _____

   _____

   _____

UNDERSTANDING KEY IDEAS

2. Identify five endocrine glands, and explain why their hormones are important to your body.

   _____

   _____

   _____

   _____

   _____

   _____

_____ 3. Hormone imbalances may cause
   a. feedback and insulin.
   b. diabetes and stunted growth.
   c. thyroid and pituitary.
   d. glucose and glycogen.

4. How do feedback mechanisms control hormone production?

   _____

   _____

   _____

   _____

Section Review *continued*

MATH SKILLS

5. One's bedtime blood-glucose level is normally 140 mg/dL. Ty's blood-glucose level is 189 mg/dL at bedtime. What percentage above 140 mg/dL is Ty's level? Show your work below.

CRITICAL THINKING

6. Making Inferences Glucose is a source of energy. Epinephrine quickly increases the blood-glucose level. Why is epinephrine important in times of stress?

_____

_____

_____

_____

_____

7. Applying Concepts The hormone glucagon is released when glucose levels fall below normal. Explain how the hormones glucagon and insulin work together to control blood-glucose levels.

_____

_____

_____

_____

_____

_____

Skills Worksheet)

Chapter Review

USING KEY TERMS

Complete each of the following sentences by choosing the correct term from the word bank.

insulin	axon	hormone
nerve	retina	neuron
reflex	central nervous system	

1. The two parts of your _____ are your brain and spinal cord.

2. Sensory receptors in the _____ detect light.

3. Epinephrine is a(n) _____ that triggers the fight-or-flight response.

4. A(n) _____ is an involuntary and almost immediate movement in response to a stimulus.

5. One hormone that helps to regulate blood-glucose levels is _____.

6. A(n) _____ is a specialized cell that receives and conducts electrical impulses.

UNDERSTANDING KEY IDEAS

_____ **7.** Which of the following has receptors for smelling?
  a. cochlea cells
  b. thermoreceptors
  c. olfactory cells
  d. optic nerve

_____ **8.** Which of the following allow you to see the world in color?
  a. cones
  b. rods
  c. lenses
  d. retinas

_____ **9.** Which of the following glands makes insulin?
  a. adrenal gland
  b. pituitary gland
  c. thyroid gland
  d. pancreas

Chapter Review *continued*

_____**10.** The peripheral nervous system does NOT include
     a. the spinal cord.   **c.** sensory receptors.
     b. axons.   **d.** motor neurons.

_____**11.** Which part of the brain regulates blood pressure?
     a. right cerebral hemisphere   **c.** cerebellum
     b. left cerebral hemisphere   **d.** medulla

_____**12.** The process in which the endocrine system, the digestive system, and the circulatory system control the level of blood glucose is an example of
     a. a reflex.
     b. an endocrine gland.
     c. the fight-or-flight response.
     d. a feedback mechanism.

13. What is the difference between the somatic nervous system and the autonomic nervous system? Why are both systems important to the body?

_____

_____

_____

14. Why is the endocrine system important to your body?

_____

15. What is the relationship between the CNS and the PNS?

_____

_____

_____

16. What is the function of the bones in the middle ear?

_____

_____

17. Describe two interactions between the endocrine system and the body that happen when a person is frightened.

_____

_____

_____

CRITICAL THINKING

18. Concept Mapping Use the following terms to create a concept map: *nervous system, spinal cord, medulla, peripheral nervous system, brain, cerebrum, central nervous system,* and *cerebellum.*

Chapter Review *continued*

19. Making Comparisons Compare a feedback mechanism with a reflex.

_____

_____

_____

_____

20. Analyzing Ideas Why is it important to have a lens that can change shape inside the eye?

_____

_____

_____

_____

21. Applying Concepts Why is it important that reflexes happen without thinking about them?

_____

_____

_____

_____

22. Predicting Consequences What would happen if your autonomic nervous system stopped working?

_____

_____

_____

_____

23. Making Comparisons How are the nervous system and the endocrine system similar? How are they different?

_____

_____

_____

_____

Chapter Review *continued*

INTERPRETING GRAPHICS
Use the diagram below to answer the questions that follow.

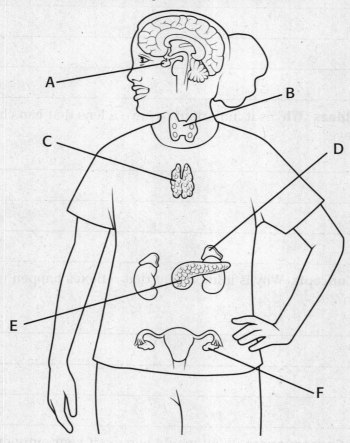

24. Which letter identifies the gland that regulates blood-glucose level?

_____

_____

25. Which letter identifies the gland that releases a hormone that stimulates the birth process?

_____

_____

26. Which letter identifies the gland that helps the body fight disease?

_____

_____

Skills Worksheet

Section Review

Animal Reproduction
USING KEY TERMS
For each pair of terms, explain how the meanings of the terms differ.

1. *internal fertilization* and *external fertilization*

_____

_____

2. *asexual reproduction* and *sexual reproduction*

_____

_____

UNDERSTANDING KEY IDEAS

_____ **3.** In humans, each egg and each sperm contain
   a. 23 chromosomes.
   b. 46 chromosomes.
   c. 69 chromosomes.
   d. 529 chromosomes.

4. List three types of asexual reproduction.

_____

_____

_____

5. How do monotremes differ from marsupials?

_____

_____

6. Describe the process of meiosis.

_____

_____

7. Are humans placental mammals, monotremes, or marsupials? Explain.

_____

_____

| Section Review *continued*

MATH SKILLS

8. Some bristlecone pine needles last 40 years. If a tree lives for 3,920 years, how many sets of needles might it grow? Show your work below.

CRITICAL THINKING

9. Making Inferences Why is reproduction as important to a bristlecone pine as it is to a fruit fly?

_____

_____

_____

_____

10. Applying Concepts Describe one advantage of internal fertilization over external fertilization.

_____

_____

_____

_____

_____

Section Review

Human Reproduction

USING KEY TERMS

1. Use the following terms in the same sentence: *uterus* and *vagina*.

_____

_____

UNDERSTANDING KEY IDEAS

2. Describe two problems of the reproductive system.

_____

_____

3. Identify the structures and functions of the male and female reproductive systems.

_____

_____

_____ 4. Identical twins happen once in 250 births. How many pairs of these twins might be at a school with 2,750 students?
  a. 1
  b. 11
  c. 22
  d. 250

MATH SKILLS

5. In one country, 7 out of 1,000 infants die before their first birthday. Convert this figure to a percentage. Is your answer greater than or less than 1%? Show your work below.

CRITICAL THINKING

6. Making Inferences What is the purpose of the menstrual cycle?

_____

_____

_____

_____

7. Applying Concepts Twins can happen when a zygote splits in two or when two eggs are fertilized. How can these two ways of twin formation explain how identical twins differ from fraternal twins?

_____

_____

_____

_____

_____

_____

8. Predicting Consequences How might cancer of the testes affect a man's ability to make sperm?

_____

_____

_____

_____

_____

Name _____ Class _____ Date _____

CRITICAL THINKING

6. Making Inferences What is the purpose of the menstrual cycle?

_____

_____

_____

_____

7. Applying Concepts Twins can happen when an embryo splits in two or when two eggs are fertilized. How can these two ways of twin formation explain how identical twins differ from fraternal twins?

_____

_____

_____

_____

_____

_____

8. Predicting Consequences How might cancer of the testes affect a man's ability to make sperm?

_____

_____

_____

_____

_____

_____

Section Review continued

8. What are five stages of human development?

_____

_____

_____

_____

_____

MATH SKILLS

9. Suppose a person is 80 years old and that puberty took place when he or she was 12 years old.

a. Calculate the percentage of the person's life that he or she spent in each of the four stages of development that follow birth. Show your work below.

b. Make a bar graph showing the percentage for each stage. Show your work below.

| **Section Review** *continued*

CRITICAL THINKING

10. Applying Concepts Why does the egg's covering change after a sperm has entered the egg?

_____

_____

_____

_____

_____

_____

11. Analyzing Ideas Do you think any one stage of development is more important than other stages? Explain your answer.

_____

_____

_____

_____

_____

_____

_____

Skills Worksheet

Chapter Review

USING KEY TERMS

For each pair of terms, explain how the meanings of the terms differ.

1. *internal fertilization* and *external fertilization*

_____

_____

2. *testes* and *ovaries*

_____

_____

3. *asexual reproduction* and *sexual reproduction*

_____

_____

4. *fertilization* and *implantation*

_____

_____

5. *umbilical cord* and *placenta*

_____

_____

UNDERSTANDING KEY IDEAS

Multiple Choice

_____ **6.** The sea star reproduces asexually by
   a. fragmentation.
   b. budding.
   c. external fertilization.
   d. internal fertilization.

_____ **7.** Which list shows in order sperm's path through the male reproductive system?
   a. testes, epididymis, urethra, vas deferens
   b. epididymis, urethra, testes, vas deferens
   c. testes, vas deferens, epididymis, urethra
   d. testes, epididymis, vas deferens, urethra

Chapter Review *continued*

_____ **8.** Identical twins are the result of
   a. an embryo splitting in two.
   b. two separate eggs being fertilized.
   c. budding in the uterus.
   d. external fertilization.

_____ **9.** If the onset of menstruation is counted as the first day of the menstrual cycle, on what day of the cycle does ovulation typically occur?
   a. 2nd day
   b. 5th day
   c. 14th day
   d. 28th day

_____ **10.** How do monotremes differ from placental mammals?
   a. Monotremes are not mammals.
   b. Monotremes have hair.
   c. Monotremes nurture their young with milk.
   d. Monotremes lay eggs.

_____ **11.** All of the following are sexually transmitted diseases EXCEPT
   a. chlamydia.
   b. AIDS.
   c. infertility.
   d. genital herpes.

_____ **12.** Where do fertilization and implantation, respectively, take place?
   a. uterus, fallopian tube
   b. fallopian tube, vagina
   c. uterus, vagina
   d. fallopian tube, uterus

Short Answer

13. Which human reproductive organs produce sperm? produce eggs?

_____

_____

_____

_____

14. Explain how the fetus gets oxygen and nutrients and how it gets rid of waste.

_____

_____

_____

15. What are four stages of human life following birth?

_____

_____

_____

_____

16. Name three problems that can affect the human reproductive system, and explain why each is a problem.

_____

_____

_____

17. Draw a diagram showing the structures of the male and female reproductive systems. Label each structure, and explain how each structure contributes to fertilization and implantation.

CRITICAL THINKING

18. Concept Mapping Use the following terms to create a concept map: *asexual reproduction, budding, external fertilization, fragmentation, reproduction, internal fertilization,* and *sexual reproduction.*

Chapter Review *continued*

19. Identifying Relationships The environment in which organisms live may change over time. For example, a wet, swampy area may gradually become a grassy area with a small pond. Explain how sexual reproduction may give species that live in a changing environment a survival advantage.

_____

_____

_____

_____

_____

_____

20. Applying Concepts What is the function of the uterus? How is this function related to the menstrual cycle?

_____

_____

_____

_____

_____

_____

21. Making Inferences In most human body cells, the 46 chromosomes are duplicated during cell division so that each new cell receives 46 chromosomes. Cells that make eggs and sperm also split and duplicate their 46 chromosomes. But then, in the process of meiosis, the two cells split again to form four cells (egg or sperm) that each have 23 chromosomes. Why is meiosis important to human reproduction and to the human species?

_____

_____

_____

_____

_____

_____

Chapter Review *continued*

INTERPRETING GRAPHICS

The following graph illustrates the cycles of the female hormone estrogen and the male hormone testosterone. The bottom line shows the estrogen level in a female over 28 days. The top line shows the testosterone level in a male over the same amount of time. Use the graph below to answer the questions that follow.

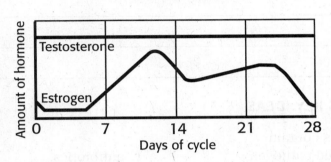

Hormone Cycles

22. What is the major difference between the levels of the two hormones over the 28 days?

_____

_____

23. What cycle do you think estrogen affects?

_____

_____

24. Why might the level of testosterone stay the same?

_____

_____

25. Do you think that the above estrogen cycle would change in a pregnant woman? Explain your answer.

_____

_____

_____

_____

_____

Name _____ Class _____ Date _____

Skills Worksheet

Section Review

Disease

USING KEY TERMS

1. In your own words, write a definition for each of the following terms:
   infectious disease, *noninfectious disease*, and *immunity*.

   _____

   _____

   _____

UNDERSTANDING KEY IDEAS

_____ 2. Vaccines contain
   a. treated pathogens.
   b. heat.
   c. antibiotics.
   d. pasteurization.

3. List five ways that you might come into contact with a pathogen.

   _____

   _____

   _____

   _____

   _____

4. Name five ways to avoid and/or fight pathogens.

   _____

   _____

   _____

   _____

MATH SKILLS

5. If 10 people with the virus each expose 25 more people to the virus, how many people will be exposed to the virus? Show your work below.

CRITICAL THINKING

6. Identifying Relationships Why might the risk of infectious disease be high in a community that has no water treatment facility?

_____

_____

_____

7. Analyzing Methods Explain what might happen if a doctor did not wear gloves when treating patients.

_____

_____

_____

8. Applying Concepts Why do vaccines for diseases in animals help prevent some illnesses in people?

_____

_____

_____

_____

Name _____ Class _____ Date _____

Skills Worksheet

Section Review

Your Body's Defenses

USING KEY TERMS

For each pair of terms, explain how the meanings of the terms differ.

1. *B cell* and *T cell*

_____

_____

2. *autoimmune disease* and *allergy*

_____

_____

UNDERSTANDING KEY IDEAS

_____ **3.** Your body's first line of defense against pathogens includes
    a. skin.   **c.** T cells.
    b. macrophages.   **d.** B cells.

4. List three ways your body defends itself against pathogens.

_____

_____

_____

5. Name three different cells in the immune system, and describe how they respond to pathogens.

_____

_____

_____

6. Describe four challenges to the immune system.

_____

_____

_____

_____

| **Section Review** *continued* |

7. What characterizes a cancer cell?

_____

_____

CRITICAL THINKING

8. Identifying Relationships Can your body make antibodies for pathogens that you have never been in contact with? Why or why not?

_____

_____

_____

_____

9. Applying Concepts If you had chickenpox at age 7, what might prevent you from getting chickenpox again at age 8?

_____

_____

_____

_____

INTERPRETING GRAPHICS

10. Look at the graph below. Over time, people with AIDS become very sick and are unable to fight off infection. Use the information in the graph below to explain why this occurs.

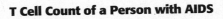

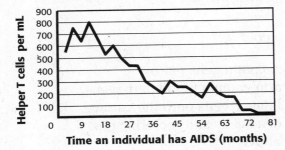

T Cell Count of a Person with AIDS

_____

_____

Skills Worksheet

Chapter Review

USING KEY TERMS

Complete each of the following sentences by choosing the correct term from the word bank.

antibody	cancer	infectious disease
B cell	noninfectious disease	T cell
pathogen	allergy	

1. A(n) _____ is caused by a pathogen.

2. Antibiotics can be used to kill a(n) _____.

3. Macrophages attract helper _____.

4. A(n) _____ binds to an antigen.

5. An immune-system overreaction to a harmless substance is a(n)

_____.

6. _____ is the unregulated growth of cells.

UNDERSTANDING KEY IDEAS

Multiple Choice

_____ **7.** Pathogens are
  a. all viruses and microorganisms.
  b. viruses and microorganisms that cause disease.
  c. noninfectious organisms.
  d. all bacteria that live in water.

_____ **8.** Which of the following is an infectious disease?
  a. allergies
  b. rheumatoid arthritis
  c. asthma
  d. a common cold

_____ **9.** The skin keeps pathogens out by
  a. staying warm enough to kill pathogens.
  b. releasing killer T cells onto the surface.
  c. shedding dead cells and secreting oils.
  d. All of the above

Chapter Review *continued*

_____10. Memory B cells
    a. kill pathogens.
    b. activate killer T cells.
    c. activate killer B cells.
    d. produce B cells that make antibodies.

_____11. A fever
    a. slows pathogen growth.
    b. helps B cells multiply faster.
    c. helps T cells multiply faster.
    d. All of the above

_____12. Macrophages
    a. make antibodies.
    b. release helper T cells.
    c. live in the gut.
    d. engulf pathogens.

Short Answer

13. Explain how macrophages start an immune response.

_____

_____

14. Describe the role of helper T cells in responding to an infection.

_____

_____

15. Name two ways that you come into contact with pathogens.

_____

_____

_____

CRITICAL THINKING

16. Concept Mapping Use the following terms to create a concept map: *macrophages, helper T cells, B cells, antibodies, antigens, killer T cells,* and *memory B cells.*

17. Identifying Relationships Why does the disappearance of helper T cells in AIDS patients damage the immune system?

_____

_____

18. Predicting Consequences Many people take fever-reducing drugs as soon as their temperature exceeds 37°C. Why might it not be a good idea to reduce a fever immediately with drugs?

_____

_____

19. Evaluating Data The risk of dying from a whooping cough vaccine is about one in 1 million. In contrast, the risk of dying from whooping cough is about one in 500. Discuss the pros and cons of this vaccination.

_____

_____

_____

INTERPRETING GRAPHICS

The graph below compares the concentration of antibodies in the blood the first time you are exposed to a pathogen with the concentration of antibodies the next time you are exposed to the pathogen. Use the graph below to answer the questions that follow.

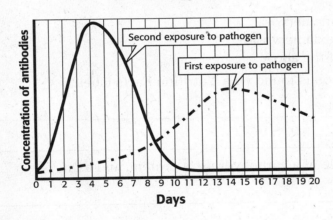

20. Are there more antibodies present during the first week of the first exposure or the first week of the second exposure? Why do you think this is so?

_____

_____

21. What is the difference in recovery time between the first exposure and second exposure? Why?

_____

_____

_____

Section Review

Good Nutrition

USING KEY TERMS

1. In your own words, write a definition for each of the following terms: *nutrient*, *mineral*, and *vitamin*.

_____

_____

_____

UNDERSTANDING KEY IDEAS

_____ **2.** Malnutrition can be caused by
    a. obesity.
    b. bulimia nervosa.
    c. anorexia nervosa.
    d. All of the above

3. What information is found on a Nutrition Facts label?

_____

_____

4. Give an example of a carbohydrate, a protein, and a fat.

_____

_____

_____

5. If vitamins and minerals do not supply energy, why are they important to a healthy diet?

_____

_____

6. How do anorexia nervosa and bulimia nervosa differ?

_____

_____

Section Review continued

7. How can someone who is obese suffer from malnutrition?

_____

_____

MATH SKILLS

8. If you eat 2,500 Cal per day and 20% are from fat, 30% are from protein, and 50% are from carbohydrates, how many Calories of each nutrient do you eat? Show your work below.

CRITICAL THINKING

9. Applying Concepts Name some of the nutrients that can be found in a glass of milk.

_____

_____

10. Identifying Relationships Explain how eating a variety of foods can help ensure good nutrition.

_____

_____

_____

11. Predicting Consequences How would your growth be affected if your diet consistently lacked important nutrients?

_____

_____

12. Applying Concepts Explain how you can use the Nutrition Facts label to choose food that is high in calcium.

_____

_____

_____

Skills Worksheet

Section Review

Risks of Alcohol and Other Drugs

USING KEY TERMS

1. In your own words, write a definition for the terms *drug*, *addiction*, and *narcotic*.

_____

_____

_____

UNDERSTANDING KEY IDEAS

_____ **2.** Which of the following products does NOT contain a drug?
  a. cola
  b. fruit juice
  c. herbal tea
  d. cough syrup

3. Describe the difference between physical and psychological dependence.

_____

_____

_____

4. What is the difference between drug use and drug abuse?

_____

_____

_____

5. How does addiction occur, and what are two consequences of drug addiction?

_____

_____

_____

6. Name two different kinds of illegal drugs, and give examples of each.

_____

_____

_____

| Section Review *continued*

MATH SKILLS

7. If 2,200 people between the ages of 16 and 20 die every year in alcohol-related car crashes, how many die every day? Show your work below.

CRITICAL THINKING

8. Analyzing Relationships How are nicotine, alcohol, heroin, and cocaine similar? How are they different?

_____

_____

_____

9. Analyzing Ideas What are two ways that a person who abuses drugs can get in trouble with the law?

_____

_____

10. Predicting Consequences How can drug abuse damage family relationships?

_____

_____

11. Making Inferences Driving a car while under the influence of drugs can put others in danger. Describe another situation in which one person's drug abuse could put other people in danger.

_____

_____

_____

Skills Worksheet

Section Review

Healthy Habits

USING KEY TERMS

Complete each of the following sentences by choosing the correct term from the word bank.

hygiene	aerobic exercise
sleep	stress

1. The science of protecting your health is called _____.

2. _____ strengthens your heart, lungs, and bones.

3. _____ is the physical and mental response to pressure.

UNDERSTANDING KEY IDEAS

_____ 4. Which of the following is important for good health?
   a. irregular exercise
   b. getting your hair cut
   c. taking care of your teeth
   d. getting plenty of sun

5. List two things you should do when calling for help in a medical emergency.

_____

_____

_____

_____

6. List three ways to stay safe when you are outside, and three ways to stay safe at home.

_____

_____

_____

_____

_____

| Section Review *continued*

7. How do seat belts and safety equipment protect you?

_____

_____

MATH SKILLS

8. It is estimated that only 65% of adults wear their seat belts. If there are 10,000 people driving in your area right now, how many of them are wearing their seat belts? Show your work below.

CRITICAL THINKING

9. Applying Concepts What situations cause you stress? What can you do to help relieve the stress you are feeling?

_____

_____

_____

_____

10. Making Inferences According to the newspaper, the temperature outside is 61°F right now. Later, it will be 90°F outside. If you and your friends want to play soccer in the park, what should you wear? What should you bring with you?

_____

_____

_____

_____

_____

Skills Worksheet)

Chapter Review

USING KEY TERMS

Complete each of the following sentences by choosing the correct term from the word bank.

nutrients   MyPyramid   addiction

drug   malnutrition

1. Carbohydrates, proteins, fats, vitamins, minerals, and water are the six

   categories of _____.

2. The _____ divides foods into six groups and gives a
   recommended number of servings for each group.

3. Both bulimia nervosa and anorexia nervosa cause _____.

4. A physical or psychological dependence on a drug can lead to

   _____.

5. A(n) _____ is any substance that causes a change in a
   person's physical or psychological state.

UNDERSTANDING KEY IDEAS

Multiple Choice

_____ **6.** Which of the following statements about drugs is true?
- **a.** A child cannot become addicted to drugs.
- **b.** Smoking just one or two cigarettes is safe for anyone.
- **c.** Alcohol is not a drug.
- **d.** Withdrawal symptoms may be painful.

_____ **7.** What does alcohol do to the central nervous system (CNS)?
- **a.** It speeds the CNS up.
- **b.** It slows the CNS down.
- **c.** It keeps the CNS regulated.
- **d.** It has no effect on the CNS.

_____ **8.** To keep your teeth healthy,
- **a.** brush your teeth as hard as you can.
- **b.** use a toothbrush until it is worn out.
- **c.** brush at least twice a day.
- **d.** floss at least once a week.

_____ **9.** According to MyPyramid, what foods should you eat most?
   a. meats
   b. milk, yogurt, and cheese
   c. fruits and vegetables
   d. bread, cereal, rice, and pasta

_____ **10.** Which of the following can help you deal with stress?
   a. ignoring your homework
   b. drinking a caffeinated drink
   c. talking to a friend
   d. watching television

_____ **11.** Tobacco use increases the risk of
   a. lung cancer.   **c.** liver damage.
   b. car accidents.   **d.** depression.

Short Answer

12. Are all narcotics illegal? Explain.

_____

_____

_____

_____

13. What are three dangers of tobacco and alcohol use?

_____

_____

_____

14. What are the three types of nutrients that provide energy in Calories, and what is the main function of each type in the body?

_____

_____

_____

15. Name two conditions that can lead to malnutrition.

_____

_____

Chapter Review *continued*

16. Explain why you should always wear safety equipment when you ride your bicycle.

_____

_____

_____

_____

CRITICAL THINKING

17. **Concept Mapping** Use the following terms to create a concept map: *carbohydrates, water, proteins, nutrients, fats, vitamins, minerals, saturated fats,* and *unsaturated fats.*

| Chapter Review *continued*

18. Applying Concepts You have recently become a vegetarian, and you worry that you are not getting enough protein. Name two foods that you could eat to get more protein.

_____

_____

19. Analyzing Ideas Your two-year-old cousin will be staying with your family. Name three things that you can do to make sure that the house is safe for a young child.

_____

_____

_____

INTERPRETING GRAPHICS

Look at the pictures below. The people in the pictures are not practicing safe habits. List the unsafe habits shown in these pictures. For each unsafe habit, tell what the corresponding safe habit is.

20.

21.

_____

_____

_____

_____